# The Historic Henry Rifle

## Oliver Winchester's Famous Civil War Repeater

### by Wiley Sword

ANDREW MOWBRAY PUBLISHERS • 54 E. SCHOOL ST. • WOONSOCKET, RI 02865 USA

LIBRARY OF CONGRESS
CATALOG CARD NO. 2002101529
    Wiley Sword
    *The Historic Henry Rifle — Oliver Winchester's Famous Civil War Repeater*
Woonsocket, R.I.: Andrew Mowbray Incorporated — Publishers
    104 pp.

ISBN: 0-931464-01-4

©2002, 2006 Wiley Sword


All rights reserved. No part of this book may be reproduced in any form or by any means without permission in writing from the author.



To order more copies of this book call 1-800-999-4697. Free catalogs available. Visit our website: www.manatarmsbooks.com

Printed in China.

This book was designed and set in type by Jo-Ann Langlois.

---

Front Cover —
   **Henry rifle, serial number 2544, associated with the 66th Illinois Infantry, Birge's W.S.S.**
                              (WILEY SWORD COLLECTION. PHOTO: KERRY BOWMAN, TROY, MICHIGAN)

Back Cover —
   *(top, right)* **C.S. Soldier. An unidentified but dapper rifleman with an early-profile Henry rifle.**
                              (RICHARD F. CARLILE COLLECTION)
   *(center, color)* **Henry rifle, serial number 6, presented to President Abraham Lincoln.**
                              (COURTESY SMITHSONIAN INST.)
   *(bottom, left)* **Unidentified Union Army 1st Sergeant with an early-production Henry rifle.**
                              (HERB PECK, JR. COLLECTION)



10   9   8   7   6   5   4   3   2

# ACKNOWLEDGMENTS

There were many who were involved in this project, and without the generous assistance of various knowledgeable collectors and students of our arms history, this work would be far less substantive. For the help of the following individuals I am particularly appreciative:

>Richard A. Baumgartner, Huntington, WV
>Nathan Bender, Cody, WY
>Doug Bennick, Orange, CA
>David Bichrest, Gray, ME
>Richard F. Carlile, Dayton, OH
>Dr. Michael R. Cunningham, Louisville, KY
>Dan Fagen, Florissant, MO
>Norm Flayderman, Fort Lauderdale, FL
>Matt Gallman, Gettysburg, PA
>Jim Gordon, Santa Fe, NM
>Mark Jaeger, West Lafayette, IN
>Howard M. Madaus, Cody, WY
>Frank Mallory, Silver Spring, MD
>Stuart C. Mowbray, Woonsocket, RI
>Herb Peck, Jr., Nashville, TN
>Les Quick, Jefferson, OR
>Dr. Thomas E. Singelyn, Grosse Pointe, MI
>Robert Zinkgraf, West Bend, WI

Special acknowledgment is given to the McCracken Research Library, Buffalo Bill Historical Center, Cody, Wyoming, for permission to publish excerpts from an original New Haven Arms Co. Ledger Book (Copies of Letters Sent). This highly valuable research document covers two periods: May 1, 1857 to March 28, 1859 (Volcanic era), and October 8, 1862 to December 12, 1863 (Henry rifle era). Nathan Bender, Curator, was most helpful with this project.

The following institutions also provided research materials, and photographs, and they are due my special thanks:

>Library of Congress, Washington, D.C.
>U.S. Army Military History Institute, Carlisle Barracks, PA
>National Archives, Washington, D.C.

# TABLE OF CONTENTS

| | | |
|---|---|---|
| **Part I** | The Henry Rifle's Inception and Sales | 7 |
| **Part II** | The U.S. Martial Henry Rifle | 41 |
| **Part III** | The Henry Rifles of Birge's Western Sharpshooters (66th Illinois Infantry) | 57 |
| **Appendix A** | Estimated Henry Production by Serial Numbers | 71 |
| **Appendix B** | U.S. Government Purchases of the Henry Rifle 1862–1865 | 73 |
| **Appendix C** | Serial Numbers of Henry Rifles Reported by the 3rd Regiment U.S. Veteran Volunteers in 1865 for Companies B, C, I, H and K | 75 |
| **Appendix D** | Partial List of Historically Identified Henry Rifles | 78 |
| **Appendix E** | Comments Regarding the Iron-Framed Henry Rifle — List of Iron-Framed Henry Rifles | 84 |
| **Appendix F** | Basic Production Configurations of the Henry Rifle | 88 |
| **Documents** | U.S. Government Trials 1864 | 90 |
| **Endnotes** | | 95 |
| **Bibliography** | | 99 |
| **Index** | | 100 |

B. Tyler Henry from Winchester Repeating Arms Co. files, Buffalo Bill Historical Center, Cody, Wyoming.

# PART I

# The Henry Rifle's Inception and Sales

The story of the introduction of the Henry rifle was told by Oliver F. Winchester, president of the New Haven Arms Company, at a time when that firm was just beginning to see favorable results following more than four years of failure and frustration. Writing on October 17, 1862, to New Haven Arms Co. stockholder E.B. Martin, Winchester outlined the progress of the company from its recent dismal status:

*Five years ago last May [1857] the company commenced manufacturing the Volcanic firearms, starting a large lot of each size, and in about 18 months thereafter began to turn out finished arms, and put them on the market. They at once found a strong prejudice against them among the dealers, and [following] a few months vain efforts to sell them, became satisfied that there were radical defects which would prevent their ever becoming a salable article. We, of course, had no way but to finish up those we had commenced, as they would have been a total loss to have dropped them. This involved a large outlay. Most of that stock has been sold at a heavy loss, and nearly all of the costly tools and machinery for making them were rendered useless. While finishing up these arms (taking about two years), we perfected and patented certain improvements (Dec. 1860), obviating the objections to the old arms, and making a perfect thing as applied to the rifle.*

*At this point we were too much exhausted, as we felt, to proceed in getting up the tools and fixtures to manufacture the [Henry] rifle, but let them lay, and took a contract to make 3,000 revolvers for a party in New York [John Walch .31 caliber 10 shot revolvers, patent of Feb. 8, 1859] calculating to clear some $8,000 upon the same (amounting [to sales of] $26,000). But when the pistols were finished, the party failed to respond. We now have them on hand and consequently the capital they cost locked up in a lawsuit.*

*Eighteen months ago [April 1861], we commenced the construction of the tools and fixtures for making our improved rifles. We have had them on the market about three months [July 1862].*

*From the commencement of our organization, till within the past three months, five years and a half, there has not been a month in which our expenditures have not exceeded our receipts.*

*Consequently we have accumulated a very large indebtedness [$77,437]. The assets of the company cannot be sold for enough to pay its indebtedness. I have repeatedly*

# The Historic Henry Rifle — Oliver Winchester's Famous Civil War Repeater

*offered to give all of my stock to the other stockholders, and relieve me from my responsibility for the company, which covers its whole indebtedness. So much for the present value of the stock, which you will see is nothing. The stock, however, has a value, but it is entirely prospective. It arises from the fact that our new rifle is a success, and will, in time, if pressed with vigor, retrieve our past losses. But to do this further aid and support will be needed from the stockholders.*[1]

Revealed in this letter were business aspects that the prosperous shirt manufacturer, Oliver Winchester, too often had failed to grasp — product and market analysis. The Volcanic arms were clearly demonstrated to be impractical based upon the prior experience of Smith & Wesson and the original Volcanic Arms Co. in utilizing the impotent and often defective "rocket ball" ammunition.

Fortunately, Oliver Winchester was a determined man who had a viable concept and the financial means to see his ideas through. New Haven Arms Company plant superintendent B. Tyler Henry's experimentation beginning in 1858 with larger caliber metallic cartridges (.44 rimfire), in the shadow of Smith & Wesson's development of the metallic cartridge for their small caliber revolvers, had provided the means to rectify the Volcanic firearms' major defects. Yet, important financial considerations were involved in developing any new arm.[2]

*The first office of the Volcanic Arms Co., on Orange Street north of Grove Street*

**Volcanic Repeating Arms Co. office.**

(BOTH PHOTOS COURTESY BUFFALO BILL HIST. CENTER, CODY, WY. OLIN CORP. WINCHESTER ARMS COLLECTION)

**Plant of New Haven Arms Co. ca.1859.**

**PART I · The Henry Rifle's Inception and Sales**

**The Walch Revolver Contract**

The rationale of Winchester and the other principals was obvious. Despite Tyler Henry's improvements, few investors were willing to gamble on the new weapon's success by pouring in additional capital during 1860. Instead, a scheme to make money on a contract to manufacture 3,000 of the uniquely designed, 10-shot Walch revolvers was utilized, the idea being that these profits would help pay for the retooling and new fixtures for Tyler Henry's rifle. With the failure of this plan due to the broken Walch contract, the New Haven Arms Co. was mired in severe financial difficulty. Indeed, it wasn't until July 1863 that about 125 of these Walch revolvers (stated to retail at $15–$18 each) were sold to an agent at $8.80 each (80 of which were returned). Bad luck seemed to insure only further ordeal, legal expenses, and perhaps lead to the dissolution of the company.[3]

**War Prospects Change Winchester's Plans**

The advent of the Civil War in April 1861 changed everything. A demand for the best arms was soon apparent, and the prospect of large-volume military sales was foremost in every sizable arms manufacturer's mind. Oliver Winchester, like any good businessman, was eager to cash in on the opportunity, and was now willing to invest further capital to do so. Orders for the requisite machinery, tools and fixtures for B. Tyler Henry's rifle were soon placed, but the delay in establishing regular production ran to about fifteen months. Winchester appears to have tried to shorten the tooling-up period, perhaps even contracting with an outside firm such as Colt's or the Arcade Malleable Iron Co. for iron frames and iron buttplates to speed initial production. Of further delay was the need to establish a separate metallic cartridge-making operation within the plant. It would be July 1, 1862, before the Henry rifle and its .44 caliber rimfire ammunition was ready for public sale.[4]

Oliver Winchester

At this point, the old question of market acceptance again became prevalent. Would the Henry rifle find ready sales? There were two markets of first importance, military and commercial. Oliver Winchester wasted little time in pursuing both prior to having finished rifles on hand.

**Early Marketing Concepts**

From the beginning, it was obvious that obtaining government military sales would take considerable time and effort due to the old-line bureaucracy and the political circumstances that were inherent. In view of the more immediate commercial prospects, Oliver Winchester planned ambitiously in the early months of 1862. A major contract was made with the firm of John W. Brown of Columbus, Ohio, to act as the New Haven Arms Company's general agent. His traveling sales agent from the Volcanic era, William C. Stanton, made good headway in the Louisville, Kentucky, region, especially with the strong pro-union editor of the *Louisville Journal*, George D. Prentice. Editor

# The Historic Henry Rifle — Oliver Winchester's Famous Civil War Repeater

Prentice received a gift Henry rifle during the last week of June 1862, evidently personally delivered by W.C. Stanton. As Prentice noted in the June 26th *Journal*, the Henry rifle "was the most beautiful and efficient rifle we ever saw." After citing various favorable reports by Washington authorities, he wrote that William C. Stanton, who was staying at the Louisville Hotel, was making his weapons available at the store of James Low & Company on 6th Street.

Prentice was so impressed with the rifle that he wrote a column about the merits of the Henry on July 14th, saying it was "the simplest, surest, and most effective weapon that we know of," and it could be fired so rapidly that one man armed with a Henry "is equal to fifteen armed with [an] ordinary gun."[5]

## The Louisville Sales Effort

Due to the powerful influence of George D. Prentice, widely known in Union circles as the "annihilator" of Rebel interests, William C. Stanton concentrated his efforts in the region, establishing dealerships with John M. Stokes & Son, and A.B. Semple & Sons, both of Louisville. By mid-August 1862, these firms, as well as the original dealer, James Low & Co., had Henry rifles for sale, and were aggressively advertising in the *Louisville Journal*.

The result was sales to various local citizens, including Judge (Major) Dan McCook, father of the "Fighting McCooks," and also to many Home Guard militiamen. Inspired by Prentice's continuing barrage of editorial endorsements in the *Journal*, whereby the Henry was advertised as "The Best [Rifle] Ever Offered to the Public," and so proficient and effective that there was "nothing left for inventors to do [in improving firearms]," sales continued to mount.[6]

In fact, so impressed was George D. Prentice, that before William C. Stanton departed, Prentice made arrangements to become the New Haven Arms Company's Louisville sales agent. Prentice purchased about $7,000 worth of rifles (i.e., 280 rifles) at a net price of $25 each, which was accepted by agent W.C. Stanton (a price O.F. Winchester later criticized as a "mistake" by Stanton). Abruptly, the Low, Stokes and Semple ads disappeared from the *Journal's* pages as of August 30, 1862. George D. Prentice was now *the* man to see if you wanted to buy a Henry, and his sales depot at the *Journal's* office offered a complete line, including ammunition.[7]

Prentice's active participation as an agent/dealer and the favorable publicity generated by his *Louisville Journal* had created a positive sales environment for the initial marketing of the Henry rifle. By mid-October 1862, the New Haven Arms Co. had produced and sold about 900 Henrys, at least 280 of which had gone to George D. Prentice. This enabled Winchester at that time to term the rifle a success.[8]

Moreover, the good sales volume seemingly would continue following the early shipments of Henry rifles to a small but enthusiastic network of dealers in the Midwestern states of Indiana, Ohio and Kentucky. Unfortunately, Oliver Winchester soon had reason to question his luck once more.

## George D. Prentice's Panic Sales

After the invasion of Kentucky by Confederate Generals Kirby Smith and Braxton Bragg beginning in mid-August and continuing through September 1862, it appeared that Louisville might be captured by the enemy or else abandoned by Union forces. According to one amazed soldier who witnessed the chaos in the city during September, the people were so "scared [that] Bragg was coming to burn the city," that "nearly all the women and children left the city and went over the river to Jeffersonville."

Ironically, editor George D. Prentice was among the most personally dispirited. Although a staunch Unionist, his "independent-minded" son, William Courtland Prentice, had hastened to join the Confederate army following the enemy's appearance in Kentucky. Only five weeks later, on September 29, young Prentice had died of wounds received at Augusta, Kentucky, as a member of the Harris Light Artillery. The distraught Prentice offered a loving if apologetic obituary in the October 2nd *Journal*, but it was only the beginning of his terrible days of despair. Unfortunately, the panic that engulfed Louisville during September 1862 came shortly after George D. Prentice had received shipments of about 280 Henry rifles, which was more than an entire month's production. Although Kentucky's crisis dissipated following the Battle of Perryville on October 8, leading to the Confederates' withdrawal, in September, during the city's turmoil, the despairing Prentice acted with poor judgment and sold his supply of Henry rifles below cost. Although his rationale was evidently explained to the New Haven Arms Co. as being intended to supply the citizenry with effective weapons at reduced prices so as to enable them to better defend their homes, Prentice's decision was disastrous. Within days, so many of these 280 Henry rifles were sold at low prices that by the 27th of September, Prentice reported only a few rifles remaining, the "whole lot on hand" being "nearly carried off."[9]

George D. Prentice

Unfortunately for the Union, at least some of these early rifles wound up in the hands of Confederates. Due to Kentucky's divided loyalties, various rifles intended for Home Guard militiamen were, instead, diverted to the enemy.

As reported by several surrendered Union officers when Clarksville, Tennessee, was captured on August 18, 1862, one company of Lt. Col. Thomas G. Woodward's 1st Kentucky Cavalry, C.S.A., was then armed "with new sixteen shooter rifles."[10] (This diverting of rifles is further suggested by several Henrys in the serial range of 100–300 bearing Confederate markings or inscriptions, including rifle no. 165, crudely inscribed: "5th Tenn. Cav., July 27, 1862.")

Of equal misfortune for Oliver Winchester was the prospect of bright success and profitable sales that seemed to have suddenly blown up in his face. Whereas the invasion of Kentucky had been an important sales boon in arming citizens and Home Guard units, the widespread, perhaps panicked sale of Henry rifles at below cost by Prentice created an immediate turmoil. Winchester's criticism of George D. Prentice, his biggest and most important early supporter, was prompt and profound.

**Dealer Complaints and Production Problems**

As to be expected, the sudden flood of Henrys on the market at below wholesale prices caused an immediate reaction among Oliver Winchester's network of small dealers, nearly all of whom were affected due to their near proximity to Louisville. Some dealerships even quit selling the Henry rifle. Oliver Winchester had to explain to various angry dealers that Prentice had "fooled them away" without due consideration of the consequences, and thus no more rifles would be sent to Louisville to his account which, in fact, would be closed.[11]

To Prentice, however, the New Haven Arms Co. president was more circumspect, his tone even being quite considerate. On October 8, 1862, Winchester mildly told Prentice that his account was closed, except for funds due on 4,000 cartridges. Indeed, Prentice continued to receive shipments of ammunition, even as the Louisville editor remained a potential problem. Weeks earlier, Winchester had asked a dealer (Child, Pratt & Fox) who was being closed out in St. Louis to send all of the New Haven Arms Co. consigned goods to Prentice. Thus, there was further worry that Prentice might soon dump more Henry rifles on the market. Even worse, in early October, there were no longer any Henry rifles for sale through a retail dealer in Louisville, which had been a hotbed of sales activity. Winchester was so upset by this dire circumstance that he queried A.B. Semple & Sons as to whether they would take a general agency there, which they later did.

As if these complex marketing difficulties weren't enough, aside from continuing protests from his dealers, Winchester had to deal with sudden production problems; the New Haven Arms Co. in mid-October 1862 was short of rifle barrels and other parts. General Agent John W. Brown in Columbus, Ohio, who supplied his network of small dealers with wholesale Henry rifles, was told the factory would be able to ship only two rifles per day against an order for fifty. There is little wonder that Oliver Winchester was willing to turn over his ownership to the other stockholders during this hectic period.[13]

## Revised Marketing Concepts

Winchester's strategy for surviving under these difficult circumstances, even while widening public knowledge of the Henry rifle, involved his refocused marketing concepts.

To dealers and general agents he expressed the same premise — sell in small lots or to individuals at retail prices, "for the express purpose of scattering our guns as much as possible." Word-of-mouth advertising from many owners, versus selling at discount to rifle clubs and the like, seemed to be the best means of profit and promoting knowledge of the rifle. Considering the very limited factory production of about 200 rifles per month, orders were already about one month ahead of production in October 1862.

The early New Haven Arms Co. discount sheet netted for the company only $28 per rifle, this being a thirty percent discount from the retail price of $40. In late October, Winchester decided to raise prices slightly to dealers ($1.40 per rifle) plus create a new retail price of $42 (without sling swivels) in order to firm up pricing at profitable levels for all.[14]

## Henry Rifle Pricing

Retail prices of the Henry rifle and its appendages beginning in November 1862 were as follows:

## PART I • The Henry Rifle's Inception and Sales

> Standard Henry rifle . . . . . . . . . . . . . $42.00
> Sling swivels, add . . . . . . . . . . . . . . . . $2.00
> Silver plated & engraved, add. . . . . . . $10.00
> Gold plated & engraved, add . . . . . . . $13.00
> Leather case, each . . . . . . . . . . . . . . $5.00
> Cartridges, .44 cal. per 1,000. . . . . . . $17.50

A circular containing a "brief description of the Henry rifle" was the only literature available in late 1862, and the New Haven Arms Co., while happy to learn of and promote the fact that the Henry rifle had earned a high reputation in Kentucky and other Western states in a few months, was hampered by a lack of production. "We hope to average about ten rifles per day after this," wrote Winchester on October 18, 1862 (they were then averaging about 8 rifles per day, six work days in a week).[15]

**Production Difficulties**

Unfortunately, production continued to proceed at the rate of about 200 rifles per month for more than a year, which created a burden not only on the factory but on dealers who were to trying to supply rifles to a receptive market. To remedy this, the New Haven Arms Co. planned on eventually arranging for higher-volume production at another facility in the event of a large government contract being secured or perhaps other large wholesale purchases.

Much of this, of course, depended upon public acceptance of the Henry as the best and most desirable rifle of its era.

**Plated and Engraved Henry Rifles**

In the beginning, in order to promote a favorable opinion of the Henry rifle, a large proportion of these arms were plated and engraved. Using in-plant facilities developed for the Volcanic arms (many pistols of which were silver-plated and engraved as a standard practice), the New Haven Arms Company offered the Henry rifle was offered silver-plated and engraved for an extra $10, or gold-plated and engraved for an additional $13.00. These arms usually featured black walnut or rosewood stocks. Numbering among the first Henrys that were manufactured, some of these engraved and plated presentation rifles were given to high government officials. Examples of the early high proportion of engraved and plated Henrys included shipments to general agent John W. Brown: October 2, 1862—10 slung, 6 silver-plated & engraved, 4 plain [unslung]; October 10, 1862—10 slung, 6 silver-plated & engraved, 1 gold-plated & engraved; week of October 18, 1862—50 plain [unslung], 20 slung, 6 silver-plated & engraved, 4 gold-plated & engraved. As late as December 1862, Oliver Winchester furnished a Mayfield, Kentucky, dealer with 20 of the 100 Henrys he ordered "finely finished." Yet, as the demand for standard Henrys increased, the prevalence of engraved and plated rifles seemed to decline. In fact, during July of 1863, when the factory was entirely out of standard Henry rifles, Oliver Winchester told a Boston client he had "a few" silver-plated Henrys remaining on hand; they were the only guns then available, and he could spare two if they wanted these. The factory's priority obviously was on turning out standard production rifles.[16]

# SIXTY SHOTS PER MINUTE

# HENRY'S PATENT
## REPEATING
# RIFLE

### The Most Effective Weapon in the World.

This Rifle can be discharged 16 times without loading or taking down from the shoulder, or even loosing aim. It is also slung in such a manner, that either on horse or on foot, it can be **Instantly Used**, without taking the strap from the shoulder.

## For a House or Sporting Arm, it has no Equal;
### IT IS ALWAYS LOADED AND ALWAYS READY.

The size now made is 44-100 inch bore, 24 inch barrel, and carries a conical ball 32 to the pound. The penetration at 100 yards is 8 inches; at 400 yards 5 inches; and it carries with force sufficient to kill at 1,000 yards.

A resolute man, armed with one of these Rifles, particularly if on horseback, **CANNOT BE CAPTURED**.

"We particularly commend it for ARMY USES, as the most effective arm for picket and vidette duty, and to all our citizens in secluded places, as a protection against guerilla attacks and robberies. A man armed with one of these Rifles, can load and discharge one shot every second, so that he is equal to a company every minute, a regiment every ten minutes, a brigade every half hour, and a division every hour."—*Louisville Journal*.

### Address  JNO. W. BROWN,
Gen'l Ag't., Columbus, Ohio,
At Rail Road Building, near the Depot.

## Presentation Arms

In the beginning, Oliver Winchester had sought to influence key officials and influential individuals with the gift of a Henry rifle. Mindful of Samuel Colt's publicity seeking methods and his success in making key connections (especially with the presentation of revolvers to the influential for favors received or anticipated), Oliver Winchester had sent a few Henry rifles as gifts, including the one to the *Louisville Journal* in Kentucky, hoping to gain favorable publicity. Further, he entered the Henry in competitions at state fairs and other public events, seeking local endorsements. Yet, the mostly disappointing results had influenced Winchester's perspective. In December 1862, even in the midst of a slowdown in orders, Winchester advised an influential dealer, "Our [low] profits don't allow us to be as liberal as we could wish in the way of presents; notwithstanding, when judiciously made, they are good investments. We will, however, in your case, add one gun to every 100 you order, without charge, to be used as presents as in your discretion it may seem best."[17]

## Political Perspectives

Political expediencies and sales difficulties continued to occupy much of Oliver Winchester's time during the first year of Henry rifle production. He was opportunistic in seeking the favor of influential dealer/clients such as Judge R.K. Williams of Mayfield, Kentucky, and his partner, Dr. W.W. Gardner, of Paducah, Kentucky. Gardner was a state senator of "high influence" who has "great facilities for selling in the army," noted Winchester, and he put much faith on his "gaining access to points not easily reached by other parties."[18]

Winchester was convinced that key political alliances, coupled with effective commercial advertising, would result in significant success in the field. Agent John W. Brown was advertising the Henry rifle as "The Most Effective Weapon in the World," and the New Haven Arms Company's extensive marketing efforts had already been noted by rival firms, or those selling competing breechloading rifles.[19]

## The Kittredge Feud

The threat posed by the Henry rifle to the powerful B. Kittredge & Co. firm of Cincinnati seemed to herald trouble of unknown proportions. In November 1862, the Kittredge company forwarded an order for 20 Henry rifles, which Oliver Winchester refused to fill. "We understand him," [his motive] wrote the aroused Winchester to agent John W. Brown on November 11. Kittredge was the principal agent for the Frank Wesson rifle, a rival metallic cartridge-firing breechloading arm. Winchester disparaged Kittredge's intent and business practices in several letters to agent Brown. Then, when Kittredge persisted in seeking to obtain Henry rifles, he wrote directly to that firm: "We have reason to believe that your interest in another rifle has led you to use your influence to injure the sale of ours by means which appear to us at least unfair, if not dishonorable." Further, wrote Winchester, "in your ambition to monopolize sales, you have frequently bought liberal lots of certain articles, getting a liberal discount on them, and then selling them at cost, or thereabouts, compelling your competitors to do the same to get rid of their goods, thus driving them out of the market; and then using the fact to squeeze the manufacturer still harder... Our purpose is to guard our business from any such influence, and our rifles from becoming a football in the market."[20]

When Kittredge once more attempted to purchase Henrys through an allied Cairo,

# The Historic Henry Rifle — Oliver Winchester's Famous Civil War Repeater

Illinois, dealer, the New Haven firm wrote directly to Kittredge, citing recent disparaging remarks reported as coming from them, and again refused any sales. "We are not willing to have imaginary defects and flaws discovered and pointed out to customers merely for the sake of foisting some other arm on the public," asserted the company's spokesman.[21]

The New Haven Arms Co. was soon attempting to track down the source of the few Henry rifles Kittredge & Co. was found to be offering for sale in the spring of 1863 (by means of serial numbers which were kept in a log book — now lost — of factory shipments). This entire matter finally came to a head following a June 8, 1863, order from Kittredge for 40 Henry rifles and ammunition, which Winchester again refused to accept. Kittredge responded with what amounted to a challenge on July 25, 1863; they proposed a trial of the Henry against the Wesson rifle. An angry Oliver Winchester left no doubt about his attitude in an August 4th letter to Kittredge. "The egotism with which you assume to have the power of setting up one gun, or putting down another, is rather amusing, and hardly excusable in you, after your failure in a similar course towards the Colt revolver." As to the trial, Winchester announced that "we shall be pleased to meet your wishes so soon as the necessary preliminaries can be arranged, and upon the result we propose to wager not less than $5,000 nor over $10,000." That ended the matter, so far as is known, and the trial was never held.[22]

## Declining Sales Lead to Aggressive Measures

Oliver Winchester's attempts to protect his new rifle's reputation and prevent unethical competition marked a new phase in his marketing concepts. By the end of 1862, a decline in sales had reduced the backlog of orders to a worrisome level. Winchester's maneuverings to develop political alliances in the commercial market and find new dealers in major cities such as St. Louis were one example of his suddenly aggressive endeavors. As Winchester noted, demand was so "spasmodic" that he couldn't predict when the company would be on firm financial footing. He told a dealer in Philadelphia that rifles were accumulating so fast at the factory that they couldn't take back the rifles the dealer wished to return. Winchester even wrote to his principal agent John W. Brown about his long silence, "we write to inquire if you are still alive," the New Haven Arms Co. president queried half in jest on December 30, 1862.[23]

While Winchester hoped for an improvement in sales based upon the advance of the Union armies ("sales will be most stirring near the scenes of active movements and hostilities," he had commented, mindful of his Louisville experiences), Brown's lack of sales and the cancellation of other orders caused the company secretary to solicit testimonials so that they could publish an advertising catalog.[24]

## The State of Kentucky Henry Rifles

Ironically, the critical publicity break Oliver Winchester had long been seeking came at that very time from a most unlikely source — George D. Prentice, the man whose below-cost sales at Louisville in September 1862 had caused the New Haven Arms Co. enormous difficulty in coping with angry dealers. Now Prentice was the key figure in an important sale. The State of Kentucky, based upon the exploits of Captain James M. Wilson, 12th [U.S.] Kentucky Cavalry, who reportedly used a privately purchased Henry rifle in repelling an attack by seven guerrillas, determined to arm Wilson's company with the Henry rifle. On October 28, 1862, an ad for recruits for Captain James M. Wilson's company, to be "armed with the Henry rifle," appeared in the *Journal*. Prentice, with important political contacts in the state, and as the Henry rifle's agent beginning about August 30, 1862, evidently had solicited the Kentucky authorities for a trial purchase of Henry rifles. Wilson, who seems to have been a personal friend and was then raising a company, was promoted to be the recipient of these arms. Beyond the enormous publicity for the rifle generated by the *Journal*, Prentice's late summer efforts were of importance when an order for 104 Henry rifles belatedly came from the state about October 20, 1862, specifically to arm Wilson's "sharpshooters."[25]

Whereas, earlier that month, Winchester had told John W. Brown he would give Prentice no more rifles, on October 21, when Prentice forwarded his order for 120 Henry rifles, the New Haven Arms Company's president virtually reinstated the Kentucky editor. Winchester, of course, cautioned him about the need to maintain established prices.

Although he at first stated it would take sixty days to produce the order, when Winchester saw the opportunity for important publicity, he expedited delivery and cited November 27 as the expected shipping date. A few days later, Prentice forwarded an amended order for a total of 200 Henry rifles to arm two full companies of the 12th Kentucky Cavalry. Oliver Winchester was much elated. J.H. Conklin, the New Haven Arms Co. secretary, told John W. Brown on November 5 that they had just taken an order for rifles for two companies of Kentuckians. "The men composing the companies are just the ones we wish to supply, as we feel it will do us a great deal of good. They are substantial farmers, and good men. They design to whip guerrillas. They will pay $34 each for the rifles." Winchester personally explained the factory's sudden change in policy by telling Prentice on the same day that they would meet his expanded order for 200 rifles, because "they are now more needed, and will be of more use in your state than in any other…" Even more patronizingly, Winchester implied that other dealers would be made to wait while Prentice's order was filled. All Prentice had to do was to insure that the price of $34 for quantities over fifty was maintained.[27]

**A Sudden Turn of Events**

Only a few weeks after receipt of this important order, the New Haven firm learned that a portion of it had been abruptly canceled. On November 24, George D. Prentice sent a telegram canceling 120 of the 200 Henry rifles based upon the Kentucky authorities' change of mind. Oliver Winchester was much upset, as all of the cartridges [24,000] and 60 Henry rifles already had been shipped to Prentice on November 20. Moreover, another 60 rifles would be ready to ship on November 26. "The course of the authorities at Frankfort appears to us much like trifling, and we think they should be held to the order they gave," grumbled Winchester. Thus, he would hold this second batch only for 10 days pending negotiation, otherwise he would release them against other orders.[28]

Two days later, November 27, 1862, the New Haven Arms Co. shipped the second batch of 60 rifles to Prentice, indicating that the Kentucky authorities had agreed to accept at least the original order for 104 for Captain Wilson's company. Indeed, by early 1863, Captain James M. Wilson's Company M, 12th Kentucky Cavalry was fully armed with 104 Henry rifles purchased through George D. Prentice. The 120 rifles shipped to Prentice for this order are estimated to be in the approximate serial number range of 650 to 800, based upon their November 1862 manufacture. The few remaining rifles (16) Prentice received beyond the Wilson order were offered in the *Journal* of December 8, his ad reading, "a very few Henry rifles and cartridges can be had at the Journal office if inquired for immediately."[29]

In an amazing turn of events, George D. Prentice had again proved of importance to the New Haven Arms Co., just when matters seemed to be at their worst.

## Defects Result in Modifications

With a reported 1,500 Henry rifles manufactured as of the end of January 1863, Oliver Winchester's firm had, in seven months, produced and sold about two hundred rifles per month. Yet, because of the scattering of rifles among a variety of customers, complaints about defects were isolated and at random. General Agent John W. Brown had returned rifles no. 324, 359, 391 and 706 for repair in early October 1862, all of which were shipped back to Brown on October 20. However, it wasn't until Captain Wilson's company began using their Henrys that certain parts defects became apparent due to the concentration of many rifles with one user. In February and March 1863, Captain Wilson sent letters reporting failures in the Henry rifle, including 25 broken "breech pins" (about a 25% failure rate).[30]

During the winter of 1862–1863, Oliver Winchester acknowledged that it was evident the Henry rifle was prone to be disabled for two reasons, both of which involved careless handling of the weapon. By elevating the spring-supported magazine plunger to the top position and then suddenly releasing it without cartridges in the magazine tube, it caused the battering of the port in the brass frame through which the cartridges passed. This "letting fly" with the plunger, warned Winchester, was a senseless act and so damaged the port that cartridges would not readily pass. Further, snapping the hammer on the firing pin bolt without a cartridge present in the chamber, frequently caused the breaking of the bolt. Another separate problem was the small twin firing pin head that slid forward in the bolt face. These pins were sometimes too small to make proper contact and explode the cartridge upon pulling the trigger.[31]

As Oliver Winchester outlined to George D. Prentice and others when these problems surfaced that winter, corrective action would be taken by enlarging the port hole in the frame, plus strengthening the bolt and enlarging the firing pins. While these modifications were to be made on all new rifles beginning about March 1863, older Henry rifles were to be supplied as needed with new bolts and firing pins. Further, a reamer was sent to several dealers to enlarge the port hole on "battered" rifles. Of course, noted Winchester in a March 19, 1863, letter to Captain Wilson, proper handling of the Henry rifle by avoiding dry firing and "letting fly" with the magazine plunger would eliminate most problems.[32]

These irksome complaints only added to difficulties reported with many Henry cartridges failing to explode, or else the powder not burning with sufficient force. Winchester found technical problems aplenty to cope with in providing a more acceptable product in the field.

## Cartridge Difficulties

Among the earliest of concerns facing the New Haven Arms Co. was the .44 caliber Henry rifle cartridge's lack of power. On November 25, 1862, Oliver Winchester admitted they had made a mistake in not developing a stronger cartridge. The use of a longer cartridge with more powder and hence more penetration had long been considered. Yet it would have reduced the capacity of the magazine to about 13 or 14 cartridges versus the present 15, and would have added length and weight to the frame. Thus they had settled on the current 26-grain cartridge, with a 216-grain bullet.[33]

John W. Brown's complaint of weak cartridges in November 1862 was originally explained by the New Haven Arms Co. as having originated when the suspected lot was made "during a cold spell." The [fulminate] paste was not dry, reported Oliver Winchester, and user complaints that some Henry cartridges were filled with tobacco, were just not true, for the cartridges "were filled by girls," he huffed. In fact, of 100 cartridges returned for evaluation by John W. Brown, all fired properly, the New Haven president claimed. The only explanation Winchester later offered for poor performance was that the tallow applied to the bullet for lubrication might melt into the powder, thus recommending that the cartridges should be kept in a cool place.[34]

Yet, when similar complaints were received from other dealers involving different lots, Winchester was forced to make a more careful investigation. He determined that, at least in some cases, the employee cartridge maker had put the grease on too hot "in order to save time," thus the tallow had saturated the powder and caused low power or else failure to fire.[35]

Winchester was much concerned about the intensifying cartridge problems, and in November 1862 he had requested Smith & Wesson, the Volcanic's old partners, to supply 100 special .44 caliber cartridges, which were longer and fitted with a larger bullet. Due to this and other experimentation, Winchester announced in January 1863 he was now making an improved cartridge with compressed powder that provided greater power.[36]

Despite these efforts, the factory's cartridge-making operations continued to be a source of much difficulty. Winchester even inquired of Crittenden & Tebbetts of Coventry, Connecticut, in mid-April 1863 about arranging among all cartridge manufacturers to make a uniform size, with an eye toward seeking a future interchangeable supply.[37]

**The Cartridge Factory Explosion**

The risks involved in storing explosives within the factory was highlighted in early May 1863 when the New Haven Arms Co. cartridge factory was accidentally "blown up." Although the rifle manufacturing operations were minimally disrupted, the cartridge factory was soon moved to a new location, and production of cartridges resumed about the first week in June. As a result of the explosion, the New Haven Arms Co. was required by a new city ordinance to purchase powder in one-pound canisters, versus the kegs they formerly used. This, of course, further complicated supply logistics and factory efficiency. By the end of September 1863, the New Haven factory was using between 40 and 50 canisters [pounds] of gunpowder per day. In late 1863, Oliver Winchester was so exasperated with cartridge-making difficulties that he planned for a plant expansion so as to be able to make 20,000 cartridges a day. Then, when the federal government seemed concerned about the performance of the existing .44 caliber Henry cartridge, Winchester once again began experimenting with an improved, higher-powered cartridge. In all, the cartridge problem wouldn't go away and, in fact, it resulted in serious difficulty during the company's attempt to sell Henry arms to the U.S. Government.[38]

# The Historic Henry Rifle — Oliver Winchester's Famous Civil War Repeater

**Larger Sales and New Planning**

Despite the many difficulties of coping with technical problems arising from a new, innovative rifle in a largely uneducated market, Winchester's firm found that one key to a successful operation was good production and delivery. Although sales had been relatively slow during the winter of 1862–1863, this proved to be a temporary aberration. By mid-April 1863, Winchester was once again reporting orders well in advance of his supply.[39] Oliver Winchester's personal efforts to get out in the field and sell his new rifle were partly responsible for this. During a trip to Kentucky in the early spring of 1863, Winchester lobbied hard with dealers for greater sales effort, sought out new dealers, and contacted satisfied clients so as to provide testimonials for a new advertising circular that he planned.[40]

Winchester's theme at the time was that ten men armed with Henry rifles were "a match for 50 armed with any other gun," and he pushed hard with newspaper advertising in Louisville and elsewhere. One would-be customer was told that "you cannot make a better investment, nor can you get an arm that will do so good execution on Indians or any other varmint." When asked about the need for a bayonet on the Henry rifle, Winchester almost laughed. The correspondent was told, "a body of men armed with these [Henry] rifles could never get within forty rods of any enemy [without their running away]."[41]

The positive effect of Winchester's activities was promptly demonstrated by a rush of orders from the field. By late April 1863, he was telling clients that orders were sixty days ahead of supply, while dealers were informed that orders were thirty days ahead. Winchester even admitted to one dealer on May 4, 1863, "we are entirely out of rifles." This situation continued to worsen, so that by mid-June 1863, the New Haven Arms Co. was reporting that they had orders for 500 rifles ahead of supply (more than two months' production).[42]

Throughout the summer, demand so exceeded production that filling small orders would require perhaps ninety days, warned the company. In fact, various orders were referred to General Agent John W. Brown to fill from his stock, if possible. Oliver Winchester informed an exporter in August 1863 that the factory had orders for ten times more Henry rifles than they could make. This situation continued to deteriorate so that when an important order arrived for sixty Henry rifles in late October 1863, Winchester replied, "We have not a finished rifle on hand... We have sixty in the works that will be finished a week from today. This is the soonest we will have any out."[43]

**Small Production Facilities Provide Big Problems**

The major problem was, of course, the limited production due to the lack of factory space at 9 Artizan Street, along with a shortage of skilled workmen, and a lack of sufficient equipment. As late as May 23, 1863, Oliver Winchester had confided that "our facilities are small, and we can make only about 200 [Henry rifles] a month."[44]

Clearly, something had to be done to increase the supply of Henry rifles. "We are doing all we can to increase our production," Winchester told John W. Brown on May 23, 1863, "but our effort will not tell for some time as we have to wait a long time to get our machines made at big prices and hire hands at extravagant wages."[45]

Winchester's determination to increase production was reflected in large equipment orders for his New Haven factory. In September 1863, Winchester reported that

PART I • *The Henry Rifle's Inception and Sales*

Unidentified Union army 1st sergeant with an early-production Henry rifle. Note the receiver-mounted rear sight. (HERB PECK, JR. COLLECTION)

during the past few weeks the company had bought $10,000 worth of new machinery to expand capacity. Unfortunately, these new (jigging) machines weren't then available, and the maker (Massachusetts Arms Co., Chicopee Falls) wouldn't commit to a delivery date.[46]

Throughout the summer of 1863, the company was compelled to listen to dealer complaints about the factory's inability to supply, and noted their suggestions for a larger plant facility. Important agents like John W. Brown were told by Oliver Winchester, "We shall go into such an armory [factory] as you have in your eye one of these days, but must creep a little longer. By and by we hope to walk, and then we shall soon be in condition to drive." The company also informed agent T.J. Albright & Sons, St. Louis, Missouri, in mid-June 1863 that, "as our orders for rifles are now some 500 ahead of our supply, you will have to be a little patient. We are hurrying things as fast as possible." Another plea from Albright brought only a perfunctory reply, "We are now getting in more machinery and hope soon to increase our facilities so as to turn out more guns."

To another potential agent in St. Louis, Horace E. Dimick & Co., the company confided in late June 1863 they weren't able to fill any new orders, but, "We are putting in a number of new machines and are in hopes to get out of the drag before long."[47]

## Transferring Rifles to Cope with Heavy Orders

In May 1863, Winchester had cautioned John W. Brown, "During the present call for [Henry] arms I think you should call in all you have out [on consignment to retail dealers] and apply them on present orders." This exchange of rifles among dealers was needed so as to cover various orders, and seemed to be the most practical expedient during the 1863 shortage. It formed the basis of the New Haven Arms Company's attempt to cope with a suddenly oversold product.[48] Even wholesale dealers, such as A.B. Semple & Son of Louisville, were asked if they had any spare Henrys to transfer them back to the general agent, and they would still receive a factory commission, just as if they had been sold through their own store.[49]

It was a strong seller's market, and Winchester told the U.S. Government on June 24, 1863, that unless they received timely formal orders, Henry rifles might not be shipped as needed, as we "cannot afford to refuse other orders upon an uncertainty." Adding to the rapidly building prospects of additional sales, European interest in the Henry rifle had been generated, and particularly large Prussian orders seemed to be forthcoming. Since favorable foreign exchange rates ("60% in our favor") added to the company's profit margin, this was represented by Oliver Winchester as a most lucrative market. Also, from California came word that standard Henry rifles were selling for $70 to $75 each, and plated weapons at $90 to $100. In all, 1863 was rife with enormous sales opportunities, if only sufficient rifles could be produced.

## Price Increases For Bigger Profits

As to be expected, Oliver Winchester did not overlook the opportunity to exploit the existing conditions so as to increase profits for his once suffering company. Small dealer discounts were the first to be altered. Some retail dealers who had been allotted twenty-five percent discounts, such as William Read & Sons, Boston, and Philip Wilson & Company, were changed to twenty percent beginning in mid-April 1863. A month later, even general agent John W. Brown was notified he was being placed on a

*PART I • The Henry Rifle's Inception and Sales*

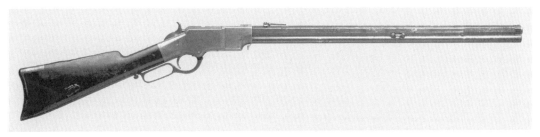

Henry rifle, serial number 729. This rifle was originally purchased during the Civil War by Mathew Wilson who served in Co. H of the 66th Ohio Volunteer Infantry. Note the sling swivel mounted to the right side of the rifle.  (DICK HARMON COLLECTION. PHOTO BY ROGER BRAHUN)

twenty percent discount. Although Winchester blamed a rise in labor costs for this increase, there is little doubt that he utilized the leverage of heavy sales to accomplish a substantial increase in the company's profitability.[51]

**Changes in the Henry Rifle**

Other more profitable changes occurred when Winchester announced on May 4, 1863, that henceforth all standard Henry rifles were to be equipped with sling swivels, since "the demand now requires slings." Of course, the added $2.00 charge (less dealer discount), standard for "slung" rifles, was unmentioned. In practice, however, the New Haven Arms Co. was circumspect, tending to supply what was ordered when the circumstances warranted. For example, in August 1863, Winchester accepted an order for 20 Henrys without slings from the important wholesale (by the case only) outlet of A.B. Semple & Sons, Louisville.[52]

Other alterations were planned in the standard Henry rifle configuration about the same time. During his trip to Kentucky and the Midwest in March 1863, Winchester had learned that many of the screw-fastened sling swivels on the Henry rifle's barrel were easily loosened during mounted service. He informed John W. Brown on March 19, immediately following his return, "we intend to sling all we make after this date in the way you suggest. I found out the mistake while out west." Since their supplier of sling swivels was the Colt's firm in nearby Hartford, Winchester promptly informed them on March 25, "For the balance of our order for swivels please omit the screws, as we shall use rivets instead." Although the factory's supply of metal screws continued to be used, rivets were intended to be utilized beginning about mid-1863. Today, many later-period Henry rifles show screw-fastened sling swivels on the barrel, indicating that the rivets were not successful.[53]

These minor changes highlighted a facet of the Henry rifle that Oliver Winchester had long considered altering, and yet was cognizant of his inability to do so.

**Special Features**

For many months, the company had been receiving letters about providing special-order features, such as globe or telescopic sights. Winchester, while often promising to provide such features in the future, found little incentive to commit the time and personnel necessary to these projects while sales were high. Only in the case of large government orders would Winchester consent to alterations in the standard Henry rifle. In a few situations, however, some minor factory modifications were made, such as a Henry for one of John W. Brown's customers in October 1862, who wanted his rifle "to pull easy on the trigger."[54]

Expediency in manufacturing the standard Henry rifle and its .44 caliber cartridges was the essence of the New Haven Arms Company's operations during 1863 and 1864. Some early appendages, such as leather rifle cases provided during mid-1862 on special orders, were eliminated as an option by the spring of 1863. Of course, the early iron-frame models, and rifles without lever-locking latches had proved less practical, and these aspects were soon altered following the Henry rifle's earliest production.[55]

**The Advertising Circular of 1863**

Following from Oliver Winchester's desire to provide meaningful advertising during the slow sales period of the winter of 1862–1863, the long-awaited company circular for the Henry rifle was available beginning about July 11, 1863.

This "description of the gun and some testimonials of its efficiency" had resulted from solicitations of Captain James M. Wilson and other members of the 12th Kentucky Cavalry, plus expert testimonials from riflemen and noted personalities. The circular was apparently prepared in June 1863 by W.S. Cleaveland of Danvers, Massachusetts, and was the source of much pride by Winchester when first supplied to prospective clients.[56]

**Company Attitudes Change**

With orders for rifles and cartridges being received in 1863 faster than the company could supply, any incentive to make concessions to the government was somewhat diminished. Sufficient funds were on hand to purchase machinery and expand operations, although allowance had to be made for a lack of skilled workmen in the labor market, and also the time delays needed to obtain additional machinery and equipment. Winchester was so optimistic about increased production in the fall of 1863, he began making promises of added deliveries despite having no finished rifles on hand.[57]

The inquiry of prominent Union officers, such as Colonel John T. Wilder, who had wanted at least 900 Henry rifles in March of 1863 for his soon-to-be-famous "Lightning Brigade," were well remembered and were a strong incentive in the fall of 1863. Whereas Winchester had to turn down Wilder for want of manufacturing capacity, which led to Wilder obtaining Spencer repeating rifles, a few weeks later he was more accommodating when Lt. Col. A.C. Ellithorpe, a prominent politician/soldier with high connections, sought to purchase 400 Henrys in April 1863. Although Oliver Winchester told Ellithorpe, "We have declined most orders for 300 to 1,000 as we cannot fill them in time without cutting off calls for smaller quantities which scatter the goods, and is more for our benefit," he promised delivery "in 60 days...if we had the order today."[58]

Winchester's optimism about increased factory production resulted in a willingness to accept orders for up to 100 more rifles per month than current production in August 1863, despite having turned down many new orders a month earlier. In fact, Winchester simultaneously increased orders for component parts and raw materials, anticipating higher production that fall.

Unfortunately, by September and October the New Haven Arms Co. was seemingly no further ahead in supplying Henrys to an ever-broadening market. For example, general agent John W. Brown was promised on October 19, 1863, only 30 rifles over 30 days. During many weeks in the fall of 1863 not a single finished rifle was in inventory, and Oliver Winchester seemed most exasperated.

Having planned for greater production by ordering new equipment and increasing supplies of component parts, Winchester was frequently disappointed in both aspects. Lathes ordered of a local firm were not good, and other equipment was not delivered as anticipated. Even more alarmingly, major suppliers such as English & Atwater of Liverpool, England (steel rifle barrels), were so delinquent on regular deliveries that Winchester complained on November 12, 1863:

*We placed with you the 2d of July last an order for 500 steel barrels, equivalent to 5,500 lbs. The last of September we received 5,721 lbs. ... August 12th we placed another order for 5,000 barrels, equivalent to 55,000 lbs., 600 barrels to be delivered Oct. 1st, and 400 barrels the first of November, and 400 the first of each month thereafter. Two shipments, 1,000 barrels are past due, and nothing has been received on this order. In the mean time we are out of work for our machinery and men, and the delay is a serious loss and inconvenience to us.*[61]

The disruption of production was a costly matter to Winchester and his firm, and with hundreds of Henry rifle orders hanging in the balance, Winchester again utilized the expediency of ordering much larger parts quantities. Five thousand finger levers were placed on order with the Arcade Malleable Iron Co., so as to improve critical parts inventories.[62]

In all, it was a most frustrating circumstance — to have more than enough orders, and too little product. Production continued to limp along at about 200 rifles per month until December 1863, when new equipment and raw material supplies enabled a slowly expanding output. Lots of from 70 to 200 rifles continued to be manufactured under the in-plant contractor system, and fortunately for the New Haven Arms Co. the increase in production came at a time when it was greatly needed to supply the U.S. Government with hundreds of Henry rifles.[63] [See Part II of this book]

**Historical Use in the Field**

Of course, the essence of the Henry rifle's great success in the market was its widespread use and endorsement by individual soldiers and officers of the Army and Navy. Since the government in 1862 and 1863 was largely uncooperative in purchasing the new Henry lever action rifle for extended trial, Winchester's firm had to rely on the individual endorsements of prominent officers and users to gain further recognition.

Due to the significant early emphasis on marketing Henry rifles in the western regions, where many dealers were active in selling the rifle, acceptance and usage continued to be much greater in the Western Union armies versus those in the East. By 1864, knowledge of the Henry rifle in the Western armies was widespread even if its use was moderate. Although all three major armies operating under Major General William Tecumseh Sherman during the Atlanta Campaign had Henrys, by far the largest user was the Army of the Tennessee, under Major General James B. McPherson. Prominent units in McPherson's command, such as the 7th, 64th and 66th Illinois, had many Henry rifles. In Major General George H. Thomas' Army of the

## The Historic Henry Rifle — Oliver Winchester's Famous Civil War Repeater

Captain Allen L. Fahnestock, Co. I, 86th Illinois Vol. Infantry, c.1863. His Henry rifle, serial no. 1348, was sold at auction in 1996. (COURTESY RICK CARLILE COLLECTION)

Cumberland, regiments with some Henrys included the 36th, 51st, 73rd, 86th and 96th Illinois, also the 40th Ohio and 40th, 57th and 58th Indiana. In the Army of the Ohio, under Major General John M. Schofield, the 65th and 97th Indiana numbered among those partially armed with Henrys.[64]

Personal accounts of the Henry's prowess proved to be of importance in this development, and among the most notable early army owners of the Henry rifle was Captain James M. Wilson of the 12th Kentucky Cavalry. As was discussed earlier, Wilson, who appears to have been a personal friend of George D. Prentice, had been instrumental in the initial purchase of 104 rifles ordered for the State of Kentucky. Oliver Winchester was most anxious to capitalize on this fine opportunity for good publicity. Repeated solicitation of Wilson and key officers in the 12th Kentucky Cavalry brought such endorsements as that supplied by Capt. Wilson on February 17, 1863: "When attacked alone by seven guerrillas I found it [the Henry rifle] to be particularly useful not only in regard to its fatal precision, but also in the number of shots held in reserve for immediate action in case of an overwhelming force. In short, I would state that, in my opinion, the Henry rifle is decidedly the best gun in the service of the United States... Give me sixty men armed with the Henry Repeating Rifle, with a sufficient quantity of cartridges, and it is not an over estimate to say that we are equal to a full regiment of men armed with muskets." Little more than eight months later, on October 27, Captain James Wilson again wrote to Winchester telling him he wanted a new Henry rifle, specially fitted for telescopic sights. Boasting that his 67 men had recently repulsed 375 of John Hunt Morgan's raiders in a 2½-hour fight, Wilson claimed his unit had killed 31 and wounded 43 of the enemy.[65]

Further adding to the luster of the 12th Cavalry's use of the Henry, a dealer reported in March 1863 how a fifteen-man scout from the 12th, armed with these rifles, were attacked by an overwhelming force and yet they "repulsed and drove from the field the [estimated] 240 assailants."[66]

Other testimonials to the effectiveness of the Henry rifle were frequently received. An officer whose men had seven Henry rifles on their 1863 trek through Arizona wrote, "In the Indian country, so great is the dependence placed upon them, that none of our men care to go on escort duty unless there is one or more of these powerful and accurate weapons in the party." John H. Ekstrand, ordnance sergeant of the 51st Illinois Infantry, reported from besieged Chattanooga, Tennessee, on November 2, 1863, that twelve Henry rifles were currently in use by that regiment, "We used your rifle in the Battle of Chickamauga with good effect, and it is undoubtedly the best gun in the service, far superior to the Spencer Rifle, or any other..." Noting that only ten rifles had been obtained from the retail dealer Bowen in Chicago, Ekstrand complained, "many more [rifles] would have been bought had I been able to get them in any place." Eckstrand was thus emphatic, "In the 51st Illinois it is many that will buy them, and the brigade and division both requested me to write to you for information and a price list..., and how many we can get, or when." We now have four months pay due us, and the boys will have the money ready to send by express to you when we can know how many we can get. Your rifle is my 'hobby' so the boys say, and I [would] like to be able to...get as many of Henry's rifles in the Army of the Cumberland, so we could drive the Rebs from Chattanooga, ...[and] get something to eat."[67]

Even Naval officers were eager to praise the Henry rifle. John S. Tennyson, pilot of the U.S. gunboat *Pittsburgh* and later the *Black Hawk*, who had purchased his Henry rifle in October 1862, called it his "pet." "I have fired about 2,000 rounds [with it] and every part of the gun is perfect yet," he wrote in June 1864. "It has been struck twice by the enemy but in no way disabled. At one time, I made them believe that there was at least twenty of us." Brigadier General Alfred W. Ellet, commanding the Mississippi Marine Brigade, found the Henry rifle so desirable he asked in January 1863 for a quotation on 1,000 of them for his men. Later, when presented with an elegant Henry rifle by C.A. Montross of the Treasury Department, Ellet wrote that "in the hands of experienced men [the Henry rifle] is the very best arm now manufactured," and "a command armed with this rifle and possessed with a proper spirit to use it, could with impunity defy ten times their number."[68]

By early 1863, with dozens of letters praising the Henry rifle on hand, the New Haven Arms Co. was convinced that they had the most effective repeating rifle on the market. Yet, what best pleased Oliver Winchester were letters such as that from Lieutenant John Brown, of Co. C, 23rd Illinois Volunteer Infantry. Writing on September 26, 1863, Brown told how he had personally purchased a Henry rifle from Adams & Co. of Wheeling, [West] Virginia, and had been so pleased with the rifle's performance that he wanted to obtain from 15 to 30 rifles for his company. Brown asserted, "It is more than probable when my company are armed with them, and their superiority established, there is but little doubt applications will be made from other companies for permission to use them, which will soon extend to other regiments, and bring them prominently before [the] government." As a result of Brown's activities, fifty Henry rifles were purchased during the winter of 1863 by the 23rd Illinois, and indeed, just as Brown predicted, the combat success of these rifles caused their lieutenant colonel, S.A. Simison, to write to the New Haven Arms Co. on February 25, 1865, asking about the cost of arming the entire battalion with the Henry.[69]

Oliver Winchester's personal faith in the Henry rifle was further rewarded when the government was finally induced in early 1864 to purchase a sufficient quantity to arm one regiment, the 1st D.C. Cavalry. This, of course, provided the means of a further advertising coup. Lafayette C. Baker's role [see Part II of this book] was most significant, and the politician/soldier supplied Oliver Winchester with glowing reports: "A thorough practical trial of your rifles in the field...has demonstrated beyond all question that it has no equal in the service," wrote Baker on December 29, 1863. "Since it has become known to the leaders of the numerous guerrilla bands that my regiment [battalion] is armed with these rifles, it is impossible for us to provoke a fight with them. Even small squads of my men have driven Mosby's and White's whole force beyond the Blue ridge, without firing a shot or drawing a saber."[70]

Further testimonials about the superiority of the Henry over other rivals were prominently utilized. Major D.S. Curtiss, 1st Maine Cavalry, wrote on January 20, 1865, "Last autumn, I with two battalions of the 1st D.C. Cavalry were transferred to the 1st Maine Cavalry regiment, armed with the Spencer seven shooters, which has given me an opportunity to test both these arms in regard to efficiency, durability and safety. I am fully satisfied that the Henry rifle is far superior in all respects, so that I would by no means use any other if it could possibly be procured. [I] believe the government would realize a great saving of life, money and time in its warfare if all the men were armed with [the] Henry rifle."[71]

Additional important praise was received from a former prisoner of war, Major Joel W. Cloudman of the 1st D.C. Cavalry, who wrote that the best evidence of the rifle's

PART I • *The Henry Rifle's Inception and Sales*

Major Daniel McCook, Sr., Paymaster U.S. Volunteers, with his Henry rifle, ca.1863. Major McCook was mortally wounded at Buffington Island, Ohio, on July 19, 1863, during John Hunt Morgan's Ohio Raid. McCook's Henry rifle, serial no. 1116, is in the Ohio Historical Society. (COURTESY OHIO HISTORICAL SOCIETY, COLUMBUS, OHIO)

# The Historic Henry Rifle — Oliver Winchester's Famous Civil War Repeater

C.S. Soldier. An unidentified but dapper rifleman with an early profile Henry rifle. (RICHARD F. CARLILE COLLECTION)

superiority came from the enemy: "I was captured last season," wrote Cloudman following his exchange, "and was for a time in Libby Prison. Several of these rifles were taken when I was, and I often heard the enemy discuss its merits. They all fear it more than any arm in our service, and I have heard them say, "Give us anything but your damned Yankee rifle that can be loaded on Sunday and fired all the week."[72]

By the end of the Civil War, there were so many good accounts of the Henry rifle's combat prowess that the weapon's special place in firearms history was assured. Inquiries from Major William S. Rowland of the elite 1st U.S. Sharpshooters and Major Franklin A. Stratton of the 11th Pennsylvania Cavalry had demonstrated the widespread appeal of the Henry rifle, while the much-sought-after word-of-mouth stories that circulated about the "sixteen shooter" spread its fame. Units such as the 10th West Virginia Mounted Infantry and especially the 64th and 66th Illinois [Birge's Western Sharpshooters] Infantry Regiments had been particularly active in privately purchasing many Henry rifles for members of their regiments. Although units such as the 32nd Illinois, 68th Ohio and 14th West Virginia Volunteers requested Henrys from the government but were denied, the private procurement and use of the Henry rifle proved of importance in the last two years of fighting. Tales by word of mouth and personal observation of the rifle's prowess, had convinced many veterans of the ultimate value of the Henry rifle, even if they had to purchase it with their own funds.[73]

**Defects Reported Result in New Designs**

Despite the great success of the Henry rifle, many of the incoming testimonial letters had asked about special features, such as improved sights, or cited various minor problems which were, of course, closely scrutinized by the company. The focus of improvements in the rifle were primarily directed in three areas: 1) the loading mechanism, 2) its weight, and 3) better ammunition.

As Major J.S. Baker of the 1st D.C. Cavalry reported in 1865, "Notwithstanding my high opinion of this arm when in the hands of dismounted men, I do not think it a suitable weapon for cavalry. I consider it too heavy; the barrel is too long for the mounted service, the coil spring used in the magazine is also liable in the cavalry to become foul with sand and mud [due to the slotted magazine tube], and this...renders the arm unserviceable." Other in-field complaints depicted the Henry's relatively delicate mechanism, the barrel's annoying tendency to overheat following rapid firing in warm weather so as to make it incapable of being held (no forestock), and the .44 caliber cartridge's lack of power. Even inquiries about better sights for the rifle were noted by the New Haven firm.

This combination of in-field praise and suggestions for improvement led Oliver Winchester to promote significant changes in the Henry rifle, at least in a conceptual form. Based upon government channeled experimentation, a new loading system was already in focus in 1864. Although eventually modified (1866) by a side-of-frame loading improvement that also allowed for a wooden forearm, early experimentation (1864–1865) was with an enclosed, muzzle-end-loading magazine tube. Shortening and lightening the barrel by use of a round versus octagonal bar stock provided for weight reduction.[76]

Today, various experimental postwar Henry rifles showing improvements have been noted by collectors. With B. Tyler Henry's departure from the New Haven Arms Co. prior to 1866, the new superintendent, Nelson King, became the primary developer of the new loading system. This former machinist from Bridgeport, Connecticut, eventually developed the famous side-loading aperture that was utilized on Winchester arms from the Model 1866 to present. Yet, postwar sales of the Henry rifle continued in the standard model until 1867, with one exception — a carbine shipped September 15, 1866, to Constantinople.[77]

**Postwar Henry Rifle Sales**

Sales were slow after the war ended in 1865 — the New Haven Arms Co. reported total 1865 sales of only 3,011 rifles, including government sales numbering 627. In fact, only 470 Henry rifles were sold during the last quarter of 1865 and the first six months of 1866, 308 of which were reportedly in 1866.[78]

Oliver Winchester now looked to the foreign market, particularly South America, for respite. Although many efforts were disappointing, it was of much importance that arms agent Thomas Emmett Addis was able to sell Benito Juarez's revolutionary

# The Historic Henry Rifle — Oliver Winchester's Famous Civil War Repeater

Mexican band about 1,000 Henry rifles and ammunition in late 1866. The $57,000 in silver coin that Addis collected in Mexico and returned to the United States involved a hazardous venture, but provided a financial godsend. Many of these Henry rifles, used in the campaign against Emperor Maximilian, are believed to have been above serial number 9000.[79]

As to be expected, factory employment had declined from 53 in late 1863 and 72 in the third quarter of 1864 to about 25 workers following the war's end. Yet Oliver Winchester wasn't discouraged. Aware that he had an improved-design, rapid-firing repeating rifle of commercial importance the world over, he quickly laid plans to capitalize on what he perceived as an existing opportunity. It was the "coming gun," Winchester asserted in his postwar advertising.[80]

## New Plans and a New Factory

Since the company was prosperous — its book value had advanced to $353,563.01 at the end of December 1866 ($182.25 per share) — this enabled plans for a larger-capacity plant and tooling for the new, improved product. Even prior to the end of the Civil War, Winchester had applied to the State of Connecticut for a new corporate charter, which was granted in July 1865, with capital stock set at the amount of $500,000. Within the year, the charter was altered to allow a change in name to the Winchester Repeating Arms Co. Winchester also divested himself of his shirt manufacturing business in 1866, selling out to his partner John M. Davies, and became Lieutenant Governor of Connecticut in 1866. In all, these events enabled Winchester to better manage his firearms business so as to provide for a new product and greater manufacturing capacity.[81]

Meanwhile, the Henry rifle continued to be manufactured in very limited quantities prior to the firm moving in late 1866 to the new Bridgeport, Connecticut, factory leased from the Wheeler & Wilson Sewing Machine Co. This plant movement undoubtedly interfered with the introduction of the new improved Henry (M1866), the first two carbines of which were shipped to H.G. Litchfield of Omaha, Nebraska, on August 31, 1867.[82]

Although the new version of the Henry rifle (later designated the Model 1866) did not come on the market until 1867, it was referred to as the Improved Henry rifle, an obvious attempt by the New Haven firm to capitalize on the well-known Henry rifle trade name.[83]

With the commercial success of the Model 1866 in the tough postwar firearms market, Winchester's calculated gamble paid off handsomely. The "Yellow Boy" Winchester Model 1866 and its subsequent Model 1873 companion arm are given the most credit for the reputation of the world-famous Winchester rifles. Yet, as the foundation of the firm's new prosperity and burgeoning firearms notoriety, the Henry rifle was truly the firearm that made it all possible.

Indeed, considering the rifle's Civil War prominence and role, the Henry was perhaps the most important rifle of all. As America's first successful repeating rifle, the Henry rifle remains a unique contribution to world firearms history.

# Color Section

*Although this book is mostly historical in its approach, rather than being a study of the physical aspects of the guns themselves, it would be impossibe to ignore the beauty and historical grandeur of these fine weapons. Therefore, we have chosen to include a small selection of color photographs that show these guns in their full glory.*

Henry Rifle and Accessories. Pictured here are a late Henry rifle with a photo of a soldier holding a Henry rifle. Lying under the rifle is a Bahn Frie bayonet, commonly believed to have been adapted to some Henry rifles. Also pictured is a box of 50 Henry cartriges, a brass cartridge box and a military-style belt and a leather carrying case. (Courtesy Les Quick. Photo by Paul Goodwin)

# The Historic Henry Rifle — Oliver Winchester's Famous Civil War Repeater

Two views of Henry rifle scabbards.
(LES QUICK COLLECTION. PAUL GOODWIN PHOTO)

COLOR SECTION

**Henry rifle #2544, associated with the 66th Illinois Infantry, Birge's W.S.S.**
(Wiley Sword Collection. Photo by Kerry bowman, Troy, Michigan)

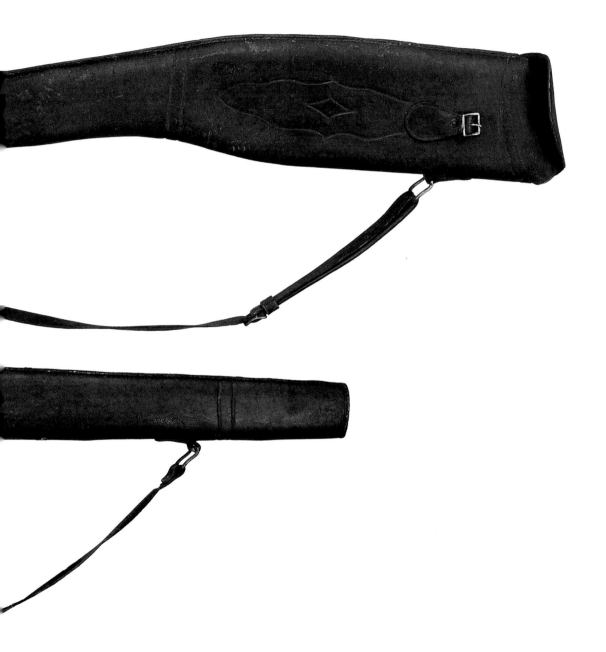

# The Historic Henry Rifle — Oliver Winchester's Famous Civil War Repeater

Two views of a presentation-grade, brass-frame Henry rifle engraved by Samuel Hodgson.   (Les Quick Collection. Paul Goodwin photo)

# COLOR SECTION

**Iron-frame Henry Rifle. A most desirable variation.**
(LES QUICK COLLECTION. PAUL GOODWIN PHOTO)

# The Historic Henry Rifle — Oliver Winchester's Famous Civil War Repeater

COLOR SECTION

Numerous views of the "Cat Gun." This exceptional Henry rifle, serial number 2363, has a brass frame and is silver plated. Engraved by Samuel Hodgson, it is one of the most elaborate Henrys known. (LES QUICK COLLECTION. PAUL GOODWIN PHOTO)

# The Historic Henry Rifle — Oliver Winchester's Famous Civil War Repeater

At the outset of regular production in mid-1862, deluxe plated and engraved Henry rifles were presented to important officials, including President Abraham Lincoln (serial number 6).

(SMITHSONIAN INST. COLLECTION)

*(below)* Henry rifle serial number 1614 carried by George W. Yerington of Company D, 66th Illinois Volunteers (Western Sharpshooters). Yerington enlisted September 27, 1861, and served at Fort Donelson, Shiloh and Corinth. He re-einlisted as a veteran on December 23, 1863, and fought through the Atlanta campaign and the march to the sea, culminating in the Battle at Bentonville, N.C. on March 21, 1865. At the time of the reenlistment, the men of the 66th purchased Henry Rifles at their own expense, which cost them about $50 each, the men owning the guns and the goverment supplying the cartridges. In February of 1864, Brigadier General G.D. Ramsay ordered from the Washington Arsenal to have made "1200 leather holsters for the Henry rifle, for issue to the 1st D.C. Cavalry, commanded by Col. L.C. Baker."

(DON TROIANI COLLECTION)

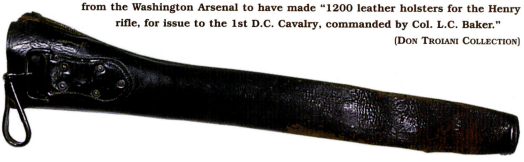

# PART II

# The U.S. Martial Henry Rifle

With the advent of the Civil War, Oliver Winchester, the enterprising businessman, couldn't long ignore the enormous potential of U.S. Government sales. Yet his problems in attempting to contract with the federal government were many.

**Early Trials and Tribulations**

As early as June 1861, Oliver Winchester had provided a handmade prototype Henry rifle for the evaluation of the Ordnance Department and particularly Brigadier General James M. Ripley, Chief of Ordnance. On the basis of what he interpreted as Ripley's favorable verbal comments, Winchester had written on June 18, 1861, to the Secretary of War, Simon Cameron, implying that Ripley had endorsed the Henry rifle. Thus, Winchester sought a "thorough test" of his rifle which had "great advantages" over all other breechloaders. Unfortunately for Oliver Winchester, it was the beginning of a strained and immensely frustrating relationship with the government bureaucracy.[1]

Ripley was provided by Cameron with a copy of Winchester's June 18th letter, and two days later the unhappy Ripley chided Winchester that, "you misunderstood me." Ripley protested that, "I had no reference to your particular arm, the beauty and ingenuity of which I did admire, but my remark was in reference to the opinion of Army officers, generally of the comparative merits of muzzle-loading and breech-loading arms as military weapons..." Moreover, asserted Ripley, "no breechloading arm, except for cavalry service, has heretofore succeeded as a military weapon."[2]

Although an unfortunate precedent, and with Ripley's austere, ultra-conservative position being evident, Oliver Winchester continued to persist in his government sales efforts. Again he submitted a prototype Henry rifle to ordnance officers, but this time for a trial by the active field army. On November 19, 1861, a successful test firing was conducted in Washington, D.C., under the direction of Captain G.D. DeRussy, 4th U.S. Artillery. Even Colonel C.P. Kingsbury, the Army of the Potomac's Chief of Ordnance endorsed the Henry rifle, saying "it would be well to purchase a number sufficient for one regiment..." Unhappily for Winchester, these favorable reports subsequently wound up on Ripley's desk.[3]

Ripley reported on December 9, 1861, to Secretary of War Cameron on the Henry and Spencer rifle "trials," plainly stating his firm convictions: "I do not consider it advisable to entertain either of the propositions for purchasing these arms," he

# The Historic Henry Rifle — Oliver Winchester's Famous Civil War Repeater

adamantly wrote. Ripley's reasoning in relation to the Henry was that the heavy weight (10 pounds when loaded), the need for a separate, special ammunition, the unproved practicality of the ammunition (particularly as to handling in the field) and even the uncertain "effect on the spiral [magazine] spring" were objections that didn't provide any important advantage over several other competing breechloaders. "The rapidity of fire" with single-shot breechloaders such as the Sharps, considered Ripley, "is sufficiently great for useful purposes without the objection to increased weight." Further, warned Ripley, "The multiplication of arms and ammunition of different kinds and patterns...is decidedly objectionable, and should, in my opinion, be stopped by the refusal to introduce any more unless upon the most full and complete evidence of their great superiority." Mindful of "the very high price asked for these arms," he had no further interest in the Henry or Spencer repeaters.[4]

This, of course, was a serious setback to Oliver Winchester. Until the New Haven Arms Company's much-delayed Henry rifle production tools, fixtures and equipment were ready in mid-1862 (about thirteen months after they were ordered in April 1861), Oliver Winchester seems to have made little further effort to secure an Army contract. Instead, his attention was turned to the bright prospect of supplying the Navy, which had a long history of buying improved weapons, including the June 4, 1861, purchase of 1,500 breechloading Sharps rifles.[5]

**Intensive Efforts to Sell to the Navy**

As production Henry rifles were nearing availability in May 1862, one of these first arms, possibly an iron-framed rifle, was submitted to Captain John A. Dahlgren at the Washington Navy Yard. Lieutenant W. Mitchell's report of the tests conducted May 16 and 17 was highly favorable. An amazing 187 shots were fired in 3 minutes and 36 seconds, and 1,040 shots were fired from the test rifle without it being disabled.[6]

Despite this successful test, and a presentation rifle having been sent to Secretary of the Navy Gideon Welles, Oliver Winchester was once again very frustrated. The Naval Bureau of Ordnance not only failed to act on the Mitchell report, but despite the unavailability of their arms, the rival Spencer rifle continued to prevail with the Navy's hierarchy. A contract for 700 Spencer repeating rifles had been initiated on June 22, 1861, following the initial test of a prototype only two weeks earlier. Since Spencer had no plant or equipment at the time of the contract, an extensive delay resulted, so much so that eighteen months elapsed (December 1862) before navy rifles were available on the Spencer contract. Thanks in large measure to the close personal friendship of Navy Secretary Gideon Welles and Charles Cheney of Connecticut (Christopher Spencer's financial backer who bought the patent and became the company's owner), by the Fall of 1862, the Spencer, not the Henry, seemed to be the Navy's favorite.[7]

All of this must have been especially infuriating to Oliver Winchester. Due to the hot prospect of sales to the Navy, around May 1862, Winchester may have ordered 200 to 300 Henry rifles manufactured with iron frames (the iron frame was less susceptible to tarnishing — i.e., the need for frequent polishing — in a salt water environment. Colt's, for example, had replaced brass with iron for triggerguards and backstraps on Model 1851 Navy revolvers sold to the U.S. Navy beginning in the 1850s).

Thus, due to crusty, old-line bureaucrats, or else political influence, the New Haven Arms Co. in 1862 seemed to be prevented from selling either to the Army or Navy.

## Oliver Winchester's Frustrations Continue

At the outset of regular production in mid-1862, deluxe plated and engraved Henry rifles were presented to important officials, including President Abraham Lincoln (serial no. 6), the new Secretary of War, Edwin Stanton (serial no. 1), and Gideon Welles, the Secretary of the Navy (serial no. 9). Unfortunately, the results were disappointing. Sales to the government were nil during 1862. Even when Oliver Winchester petitioned Secretary of War Stanton in August 1862 to allow military companies privately purchasing Henry rifles to retain them in the field, Stanton had his assistant reply "in the negative." "Great inconvenience has resulted from promises heretofore given in other cases to furnish companies of troops with special arms," sternly wrote Assistant Secretary of War Peter H. Watson. Only if Winchester would choose to arm and equip an entire regiment "at your own expense," would the government accept this arrangement.[8]

Oliver F. Winchester disgustedly confided to Brigadier General Alfred W. Ellet on January 31, 1863, that thus far "but 1,500 of Henry's Rifles have yet been made and sold; none to the government, or to any organized body, except 104 recently furnished to the State of Kentucky." Indeed, despite good reports from correspondents and satisfied owners, Winchester could but lament, "Most of the [rifles] have been sold to individual officers of the army and navy."[9] Oliver Winchester was so frustrated with the Ordnance Department, and Ripley in particular, that he prepared a long, accusatory letter about the lack of consideration given to the Henry rifle, which was published in *The Scientific American* of March 7, 1863, under the initials "O.F.W." It reads as follows:

> ...General Ripley, Chief of the Ordnance Department at Washington, opposes their introduction into the army, and has recently said that he prefers the old flintlock musket to any of the modern, improved firearms, and that he believes that nine-tenths of the army's officers will agree with him...the only reason, or excuse rather, that we have ever heard against the use in the army of arms susceptible of such rapidity of loading is that the troops would waste the ammunition... The saving of life does not appear an element worthy of consideration in this connection. Yet this is West Point opinion, the deductions of West Point science. Are these results worth their cost to the country?

## Oliver Winchester Revises His Strategy

Government sales, of course, were too important to dismiss as unavailable despite the resistance of men such as Ripley. Although the methods he contemplated were at odds with his conservative concepts, Oliver Winchester again sought to secure favorable notice and influence politically important officials. Having learned a hard lesson with Ripley and Welles, and after rethinking the basis of his company's unpromising political relationship with the government, Winchester sought to obtain direct internal leverage. In early 1863, he again solicited the favor of Colonel Lafayette C. Baker, the erstwhile "chief of detectives," who had been appointed provost marshal in Washington, D.C. Most importantly to Winchester, Baker reported directly to Edwin Stanton, Secretary of War, and was a War Department insider. The opportunistic Baker was not above lining his pockets with tendered greenbacks, and his reputation as a key Stanton cohort added to the political intrigue involved in Winchester's attempt to sell Henry rifles to the federal government.[10]

# The Historic Henry Rifle — Oliver Winchester's Famous Civil War Repeater

In March 1862, a test firing of a prototype or one of the earliest production Henry rifles in Washington, D.C., had been arranged with Baker and a few War Department "military and scientific experts." The results had been reported by Winchester's agent/promoter George D. Prentice in the *Louisville Journal* of June 26, 1862: "thirty shots were fired [by Baker] in a minute and one-third, at a distance of 100 yards, and the thirty bullet holes in the target could be covered by a piece of paper six inches square."[11]

## Initial U.S. Government Orders

Winchester's path to Stanton through Baker paid off handsomely. In fact, the New Haven Arms Company's initial breakthrough on U.S. government sales abruptly occurred during mid-1863, nearly a year after the Henry's public introduction.

Apparently, through Baker's influence, a formal War Department order via Chief of Ordnance Ripley was placed with the New Haven firm for Henry rifles in mid-June 1863. Although as early as May 23, Winchester anticipated an order for 500, Baker had forwarded his requisition through the War Department, where only 240 Henrys were approved for purchase. Baker then directly communicated the 240 quantity to Winchester, and the order was routed to the Ordnance Department for placement. Yet, as might be expected of the disapproving Ripley, he cut the reduced 240 order in half before sending it on to New Haven. On June 24, 1863, Oliver Winchester delicately raised the question of quantity in a letter to Ripley, saying, "Shall we regard his [Baker's] order [for 240] as sufficient, or wait your [revised] order [increasing the specified 120 quantity to 240] before further shipment[?]"[12]

Ironically, whereas six months earlier the New Haven Arms Co. had a surplus of Henry rifles, by mid-June 1863, Winchester was now so short of rifles that he was unable to fill from factory stock even the government's reduced order of 120. On June 20, 1863, the New Haven Arms Co. factory shipped 80 Henry rifles via Adams Express Co. to the Washington, D.C. Arsenal, c/o Col. George D. Ramsay. These 80 rifles were not formally inspected and are believed to have been randomly in the 2100 to 2300 serial number range. In order to rapidly complete this important order, Winchester solicited the balance remaining (40 rifles) from his general agent dealer, John W. Brown, in Columbus, Ohio. Brown supplied 50 Henry rifles, and shipped them directly to Col. Lafayette Baker in Washington, D.C., on June 19, without going through Ordnance Department or arsenal routing. From the evidence noted in martial Henry rifle listings among the 3rd U.S. Veteran Volunteer Infantry in 1865, it appears these 50 Henrys were randomly in the 1300–2100 "commercial" serial number range.[13]

Although ten more rifles had been supplied (a total of 130) versus the formal order for 120, Oliver Winchester craftily told Ripley that, "It will take us two to four [more] weeks to furnish the remaining 110 [rifles on the Baker-origin order for 240]." As such, the very slow production rate of between 8 and 10 rifles per day (about 200 per month) remained a millstone around the neck of the New Haven Arms Co. for months to come. Winchester had difficulty meeting the delivery dates he specified for the final 110 Henrys, and a month later, on July 21, Ripley queried Ramsay at the Washington Arsenal if he had received any more of these arms. Ramsay replied the next day that he had received only the 80 shipped directly by the New Haven Arms Co. Fifty other rifles had gone directly to L.C. Baker, noted Ramsay. Ironically, on July 21, 1863, the balance of the government's [Baker's] order was finally shipped by the factory to L.C.

## PART II • The U.S. Martial Henry Rifle

Baker. On July 25, the government acknowledged receipt of the New Haven Arms Company's invoice for a quantity of 241 Henry rifles (total of $8,676 or $36 each). Again, from recorded martial Henry rifle serial numbers on hand in the 3rd Veteran Volunteers and elsewhere, it would appear that the final 110 rifles were in the 2300–2600 serial range. The 240 total "Baker" order had now been completed.[14]

### The Henry Rifle and the 1st D.C. Cavalry

All of these rifles were intended for the direct use of Lafayette C. Baker's unique cavalry battalion, first known as Baker's Mounted Rangers (later the 1st District of Columbia Cavalry). Baker, in June 1863, had been authorized to raise a special force of four companies, which was intended for police and provost duty in the D.C. district. Baker and his unit thus came only under the direct orders of Stanton in the war department. Due to the prominence of Confederate partisan rangers like John S. Mosby in the district, it was expected Baker would effectively prevent embarrassing incursions such as the capture by Mosby in March 1863 of Union Brigadier General Edwin H. Stoughton near Washington.

An unidentified trooper of the 1st D.C. Cavalry, fully equipped for the fray with his "sixteen-shooter." (RICHARD F. CARLILE COLLECTION)

# The Historic Henry Rifle — Oliver Winchester's Famous Civil War Repeater

Essentially a Mosby targeted unit, the Mounted Rangers were soon designated the 1st D.C. Cavalry, and additional troops were planned (to be recruited in Maine) so as to raise the battalion to regimental strength. Probably through Baker, Winchester appears to have quickly been informed of the expanded 1st D.C. unit, and anticipated an increase in his 240 rifle order by perhaps another 1,000 rifles (allowing for 12 companies of 100 men each).[15]

## The Prospect of Larger Orders Leads to Big Plans

An enthusiastic Oliver Winchester was by this time considering making special-order Henry rifles for the U.S. government. On June 12, 1863, he wrote Commissioners Joel Hayden, Col. William R. Lee and Col. John Kurtz about his proposal to make a cavalry model (short barrel), or picket-size and sentry-size Henry (longer barrel). In forwarding a sample of the standard 24-inch barrel Henry rifle, Winchester stated he would even be willing to make a .50 caliber Henry with a 30-inch barrel, for $35 each, if given an order for a minimum of 2,000 (18 months required for first delivery).[16]

Accordingly, Oliver Winchester seemed to anticipate receiving large government orders that summer and fall. In July 1863, he told a correspondent he was preparing to make a carbine size and a longer, infantry size. New production equipment was ordered, and experimentation on cartridge performance and a larger caliber was begun. Two Henrys with special globe sights were prepared for trial in early September 1863 and sent to P.H. Watson, the Assistant Secretary of War.[17]

Then, in mid-September, came further good news. Ripley was replaced as the Chief of Ordnance, and the new acting chief, Colonel George D. Ramsay, sent an order on September 18, 1863, for a sample of "Henry's Carbine." Although Winchester forwarded a standard rifle (@$45.75), this being the only Henry then available, he told Ramsay that if carbines were wanted, the Henry could be made with a 19¼-inch barrel, and would still carry twelve cartridges in the magazine.[18]

## Additional Henry Rifles for the 1st D.C. Cavalry

When Lafayette C. Baker's urgent telegram arrived on October 26, 1863, ordering 60 additional Henry rifles (intended for the initial expanded company — Co. D — of the 1st D.C. Cavalry, organized in Maine, which left the state October 22 for Washington, D.C.), Oliver Winchester was much embarrassed. The factory was completely out of finished rifles. Sixty were in the works, confided Winchester, but it would be a week before they were ready. As a result, it appears that these rifles were shipped directly to Baker in Washington, D.C., by John W. Brown, the Columbus, Ohio, general agent. These 60 total Henry rifles, delivered in October of 1863, were probably in the commercial serial number range, 1300–2999. Indeed, the 1865 presence of U.S.-owned rifles no. 1392, 1491, 1492, 1592, 1732, 1798, 1804, etc., in the hands of the 3rd Regiment U.S. Veteran Volunteers, suggests that these were rifles originally shipped from the factory to John W. Brown, General Agent, Columbus, Ohio, which were in turn supplied by Brown in 1863 (50+60=110) for various Baker/government orders on behalf of the New Haven Arms Co. However, since Baker's October 1863 special order was not approved by proper authority, the Ordnance Department subsequently warned Winchester that, in the future, purchases that were made by Col. Baker or anyone other than the Ordnance Department or the Secretary of War would not be paid for.[19]

## PART II • The U.S. Martial Henry Rifle

**The Henry Carbine Project**

Oliver Winchester remained undaunted. Aware of the government's burgeoning interest in a carbine-sized Henry, he proceeded to secure what he hoped would be a bonanza of government orders. In mid-November 1863, Winchester visited Washington, D.C., to see various government officials and came away highly encouraged. On November 17, following his return, Oliver Winchester informed Assistant Secretary of War Watson that for his firm to manufacture the Henry carbine in large numbers he would have to obtain a new production facility. It would take 18 months to build a new plant and begin deliveries. Accordingly, Winchester asked if the armory at Bridgeport, Connecticut, which Dwight & Chapin had conditionally sold to Merwin & Bray to make Ballard carbines for the government, could be made available due to the Henry's higher priority. If so, Henry carbines might be produced for the Army in six months. Meanwhile, Winchester promised to begin experiments to improve the power of the current .44 caliber cartridge, and would furnish a sample carbine for "verification" of results.[20]

Within a few days, Oliver Winchester was even more optimistic. Following a visit with E.K. Root at Colt's, he obtained their cooperation for contractual manufacture of the Henry carbine at the Colt manufactory, provided sufficient volume was generated. Indeed, on November 27, 1863, Oliver Winchester offered a formal proposal to Asst. Secretary Watson, offering three options: 1) to make 40,000 Henry carbines at Colt's for $26 each, initial delivery 6 months from the date of order, with 200 per day thereafter; 2) 20,000 Henry carbines made at Colt's for $27 each, initial delivery 6 months with 100 per day thereafter; or 3) 10,000 Henry carbines made at Bridgeport for $28 each, initial delivery in 8 months, with 50 per day thereafter. All these carbines would be of the usual "cavalry" length barrel, and of the same quality as the model "we propose to make." Moreover, these Henry carbines would carry a longer .44 caliber cartridge (case 1⅗ inches) with 30 grains of powder (versus the current 26 grains). So as to minimize the loss of cartridge capacity in the magazine, Winchester already had a redesign in mind to replace the existing Henry rifle slotted magazine and plunger, which was the cause of some trouble in the field. Accordingly, on November 30, Winchester inquired of Remington & Sons about making fully enclosed magazine tubes from rolled sheet steel versus solid stock. The idea was to incorporate a plunger tube containing a spring into an open-ended tubular magazine housing under the barrel. This design, in fact, was incorporated into the sample carbine ordered manufactured for the government's evaluation (see pg. 92 for photos of this test rifle).[21]

On December 5, 1863, Winchester wrote to Colt's informing them that a model Henry carbine was being made, and when completed, they would promptly forward it to Colt's for a better estimate of the manufacturing costs.[22]

**The Block Number U.S. Martial Henry Rifles**

Meanwhile, the more immediate matter of L.C. Baker's standard Henry rifle order to equip the 1st D.C. Cavalry's new companies was in present focus. As such, an anomaly in Henry rifle production may have occurred about January 1864. Due to the existing circumstances with various government orders in prospect, it is believed the factory had earlier set aside the specific serial number range of 3000 to 4000 for the L.C. Baker order. Why? Because, as was demonstrated by the New Haven Arms Company's correspondence with the War Department, the government was very active in evaluating the Henry for large purchases, and quality assurance was to be of primary importance. Furthermore, formal U.S. Ordnance Department inspection was demanded, a procedure not performed under the original 240 rifle contract (hence the reason there were no ordnance inspector's marks on the first 240). Having a definitive means of keeping track of these "army" Henrys (by block serial number) would make for better production quality control and inspection accountability under the "lot" production operations that scattered rifle parts throughout the factory. This procedure was of further significance in view of the anticipated manufacture of Henry carbines for the Army.[23]

Based upon Baker's requisition, Brigadier General George D. Ramsay, the new Chief of Ordnance, sent a formal order on December 30, 1863, to the New Haven Arms Co. for an additional 800 Henry rifles at $36 each (so as to arm the expanded 1st D.C. Cavalry Regiment, whose new companies were then forming in Maine). Since Oliver Winchester had been long anticipating this order, on January 4, 1864, he had 200 finished Henry rifles, which he offered to deliver a week later. The balance of 600 rifles he proposed to deliver at the rate of 100 per week thereafter. Because the factory's full production capacity had remained at about 200 plus rifles per month, this suggests that many of the parts for the remaining 600 rifles were then in process, or at least materials were on hand so as to provide an expanded factory production capability. (Assuming that an "army" block of serial numbers 3000 to 4000 had been set aside in the fall of 1863, production of commercial guns [in the 4000 serial number range] may have resulted in some out of numerical sequence factory shipments during the fourth quarter of 1863.)[24]

Indeed, having planned ahead for Baker's 1st D.C. Cavalry 800-rifle order, enabled Oliver Winchester's firm to complete delivery of the entire quantity by March 9, 1864. Such would have been unlikely without prior arrangements and preproduction.[25]

Yet, as Oliver Winchester learned during his attempts to efficiently satisfy this 800 government order, dealing with a reluctant and all-but-uninformed Ordnance Department had serious drawbacks. Soon there were many difficulties that greatly exasperated Winchester and his colleagues. A letter by a disgruntled New Haven Arms Co. employee to the manager of another firm seeking employment there reveals that as early as mid-January 1864 a U.S. Ordnance inspector was at the Henry plant insisting on changes. The employee was evidently paid as a subcontractor by the piece and was angered by the anticipated delay involved, saying "it will be some time before I will have anything to do."

According to the employee, "the work was all wrong, so they have got to make an alteration in their tools." The only major change in the U.S. martial Henry rifle from 1863 standard production involved the elimination of the dovetail slotted frame as an option for the rear sight. The frame was made solid in the 1864 army production that began about serial no. 3000 (this change then became permanent on all Henry production). Although perhaps an irritant, this alteration did not take long to implement.[26] As a rule, U.S. Martial Henry rifles were furnished without sling swivels due to the added cost, although some appear to have been so equipped.[26]

## U.S. Martial Henry Rifle Inspection

On January 28, 1864, "Mr. Rice" [Ordnance Dept. sub-inspector] arrived at the New Haven Arms Co. plant to inspect the standard Henry rifles on government order.

> HEAD QUARTERS
> DEFENCES OF NORFOLK AND PORTSMOUTH,
> PORTSMOUTH, VA.,
>
> May 17 1864.
> 10 30 P.m.
>
> Colonel
>
> The Genl Comdg directs that you will have your Regiment in readiness to form on short notice
>
> Your ammunition has been sent for to Fort Monroe
>
> You will have a sufficient detail in readiness to come into town & unload it as soon as it arrives
>
> I am Colonel
> Very Resp
> Your obt servt
> S L McHenry
> Capt & AAG
>
> Lt Col Conger
> 1st DC Cavalry

Manuscript order, May 17, 1864, to Lt. Col. E.J. Conger, 1st D.C. Cavalry, directing him to immediately prepare for combat service, as their [Henry rifle] ammunition was en route. The 1st D.C. Cavalry was the only regiment in the Union army completely supplied with Henry rifles by government purchase. (COLLECTION OF WILEY SWORD)

## The Historic Henry Rifle — Oliver Winchester's Famous Civil War Repeater

At this point, 400 "army" Henry rifles were already assembled, with others in various stages of advanced assembly. However, Rice wanted these arms disassembled so as to properly inspect each part, per Ordnance Department standards.

Oliver Winchester was obviously upset, and fired off a letter to Rice's boss, Col. William A. Thornton (Chief Inspector of Contract Arms, headquarters in New York), stating his anxiety to avoid "the delay and expense of this." Specifically mentioned was the difficulty of "getting competent men to reassemble [the rifles]."[27]

Thornton's reply was shocking. On January 29, Colonel Thornton sternly advised the New Haven Arms Co. that, "Your rifles are to be inspected in detail, after the manner of all arms received by the government." Moreover, specific appendages would be required for each rifle: a cone wrench and screwdriver, a tompion, a wiper, a spring vice, a tumbler punch or trigger pin, and an extra cone.[28]

Oliver Winchester wasted no time in protesting. On January 30, he told Thornton that previously the Army's Henry rifles "have been received without inspection, [and were supplied] from stock in process." "We did not anticipate such a [detailed] inspection as is now ordered," continued Winchester. "The expense and delay to us is serious." Although he admitted they would "submit if unavoidable," the company would require additional time to procure more workmen. Moreover, as to the appendages, Winchester wrote the same day to Thornton's superior, Chief of Ordnance Ramsay, that the specified extras were "of no use with the [Henry] rifle." It was obviously a mistake and "unfair if not unjust" to require the supply of standard percussion rifle musket appendages with a repeating rifle that fired metallic cartridges. Further, since 400 Henrys were now ready for shipment, and as "we are required to strip them and reassemble them at our own expense," this would cause a lengthy delay in delivery.[29]

A few days later, Winchester outlined to Ramsay what he would be willing to do. Despite "the impression that only a general or snap inspection was usual or required," he would waive his objection and the rifles would be disassembled for inspection, providing additional time was allowed "to procure additional men to strip and reassemble the guns." However, on the matter of appendages, Winchester would not supply anything more than a cleaning rod for each rifle, and "one punch and one or more screw drivers...if required." Otherwise, "as we have orders waiting for the rifles if not wanted by the government," he would not fill the order.[30]

This critical matter was finally resolved by a tacit compromise. Colonel Thornton was instructed by Ramsay on February 9 to inspect the Henry rifles "in a way that will fully satisfy him that they are fit for service," but "not to require with your rifles any appendages but such as he deemed absolutely necessary." Some flexibility thus having been orchestrated, it appears that under Thornton's orders the Henry rifles were not fully disassembled, but were inspected in detail by ordnance sub-inspector Charles G. Chapman during late February 1864. Moreover, only the normal cleaning rod and screwdrivers and a punch were supplied.[31]

The 800 Henry rifles, serial range of 3000–4000, bearing Chapman's (C.G.C.) initials on the right side of the barrel at the frame, and (C.G.C.) cartouche on the stock, were then shipped to the Washington, D.C. Arsenal. A total of 783 of these Henrys were issued to the 1st D.C. Cavalry during March 1864.[32]

### The Henry Rifle in Combat with the 1st D.C. Cavalry

Despite its original local provost purpose, the 1st D.C. Cavalry was transferred to the Virginia peninsula during the spring of 1864, due to the need for combat-ready cavalry. Rushed into active service during May, so many of the regiment's Henry rifles were subsequently lost that on August 3 (after four months of active service) the regiment reported only 524 on hand. Brigade commander August V. Kautz was so upset by this that he wrote a memo to the Ordnance Department stating his objection to "the culpable waste and loss of these arms." The Henry rifle was "of an excellent character," noted Kautz, "[and is] certainly of the most recent and expensive model."[33]

Despite Kautz's concern, the 1st D.C. Cavalry continued to lose Henrys at a rapid rate, including heavy battle losses at Stony Creek, Virginia, June 24, 1864, and Sycamore Church on September 15. When seven 1st D.C. Cavalry companies [D, F, G, H, I, K and L] were transferred to the 1st Maine Cavalry on August 27, 1864, their Henry rifles went with them and were used by that unit.[34]

In all, a total of 1,100 Henry rifles were procured specifically for the 1st D.C. Cavalry (240 + 60 + 800), the first 300 of which were randomly in the commercial serial number range of 1300 to 3000, and the final 800 ordered were Ordnance Department-inspected Henrys in the 3000 to 4000 serial number range.

At the end of the war, a total of 62 Henry rifles were sold to discharged soldiers from the 1st Maine Cavalry, and 65 to the remaining battalion of the 1st D.C. Cavalry. Some of the other 1st D.C. Cavalry procured Henrys (both those that remained unissued at the Washington, D.C. Arsenal and turn-ins from the 1st D.C. Cavalry or 1st Maine Cavalry) may have been issued to Company F of the 97th Indiana Infantry Volunteers. Beyond this, some of these turn-ins and/or "arsenal inventory" Henrys helped arm a regiment of veteran volunteers that were organized in February and March 1865.[35]

**Late War Prospects for the Henry Carbine**

Although his eyes were opened by the frustrating experience of dealing with the government during the January 1864 inspection controversy, Oliver Winchester remained optimistic about receiving large government orders for the Henry carbine. Since the Henry rifle's old nemesis, Brigadier General James W. Ripley, had been replaced by Col. George D. Ramsay in mid-September 1863 as Chief of Ordnance, Winchester was hopeful that his relationship with Baker, plus Stanton's tacit endorsement, might result in large government orders.

Although the December 1863 special-order sample Henry carbine was delayed, this uniquely designed tubular magazine loading weapon was given to Major A.B. Dyer of the Ordnance Department on or about January 21, 1864, for evaluation. The New Haven Arms Co. had received the finished carbine only one day prior to submitting it to the Ordnance Deptartment and only briefly test fired the weapon (a total of six cartridges were expended). Moreover, an experimental new, long .44 cartridge, supplied by Leets & Co. of Springfield, Connecticut, with 37 grains of powder was used in this carbine.[36]

As evaluated by newly appointed Lieutenant William S. Smoot and armory superintendent Lucian C. Allin, the Henry carbine failed to pass the government's test. The formal report, forwarded to Oliver Winchester by Major Alexander B. Dyer, commander of the Springfield Armory, on April 2, 1864, stated that the Henry carbine had failed, becoming unserviceable three times during the trial: 1) the internal parts were easily "disarranged," 2) a spring had broken, and 3) various cartridges had burst, necessitat-

# The Historic Henry Rifle — Oliver Winchester's Famous Civil War Repeater

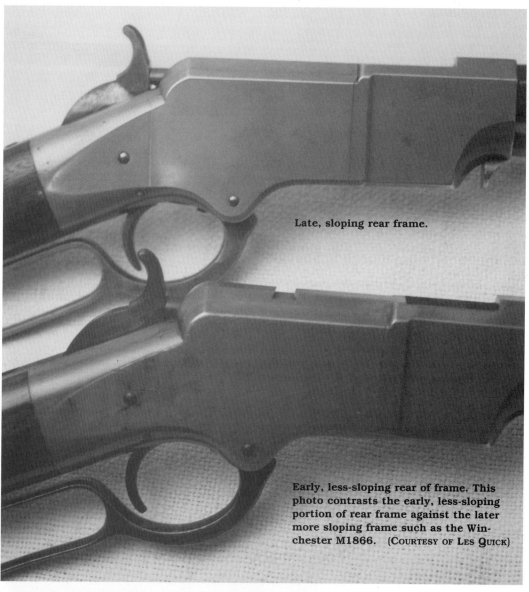

Late, sloping rear frame.

Early, less-sloping rear of frame. This photo contrasts the early, less-sloping portion of rear frame against the later more sloping frame such as the Winchester M1866. (Courtesy of Les Quick)

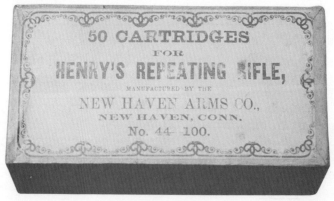

Box of Henry cartridges manufactured by New Haven Arms Company. (Courtesy of Les Quick)

ing taking the gun apart to remove the casing. Further, the Henry carbine was defective in that it had no half-cock (safety), and the mechanism was too delicate and expensive. Otherwise, the separate magazine cartridge tube plunger was likely to get lost (rendering the arm inoperable as a repeater), and the carbine's weight of 8 pounds 5 ounces was objectionable.[37]

Oliver Winchester was greatly distressed by the Army's adverse report and strongly objected to it in a letter to Dyer dated April 20. His arguments were both broad and specific; that the arm was new and largely untried, and that minor defects such as no half-cock notch safety could be easily remedied. Further, Winchester stated that the primary difficulties were with the ammunition, not the carbine. In all, asserted Oliver Winchester, the fault had been his "in not thoroughly testing & correcting such weak points...before submitting it to trial." "I mistook the spirit and manner in which I supposed such trials [were] to be conducted," he further protested. "I supposed it was the interest of the government to foster and encourage every effort to improve and perfect weapons of self defense, and if a gun had one desirable feature in it, such as this has — that of loading with great rapidity —

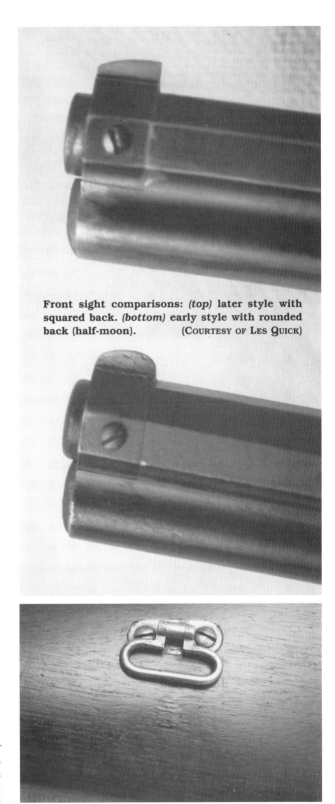

**Front sight comparisons:** *(top)* **later style with squared back.** *(bottom)* **early style with rounded back (half-moon).** (COURTESY OF LES QUICK)

**Rifle stock illustrating the infamous "bump," which on this rifle is located slightly below the swivel.** (COURTESY OF LES QUICK)

# The Historic Henry Rifle — Oliver Winchester's Famous Civil War Repeater

that feature would receive some notice, and with some encouragement to remedy the minor defects that might appear."

Notwithstanding the evaluators' criticisms, there was not one "single sound exception or rejection" to this arm, asserted Winchester, "which cannot be easily and completely removed." Further, noting that although the report was marked "final," Winchester stated that this ordnance report was an "injustice," and that Dyer should consider the nature of these objections and proceed with further evaluations.[38]

Unfortunately, much as Winchester anticipated, the formal report quickly produced a "wrong" effect. In fact, it was now apparent that Brigadier General Ramsay, the new Chief of Ordnance, favored the Spencer rifle. In April 1864, Ramsay termed the Henry rifle too expensive and "too delicate for service in its present form." No large-volume government orders would be forthcoming, and Oliver Winchester was greatly upset.[39]

Because of the anticipated government orders beyond the already high commercial demand for Henry arms, and in view of the factory's very limited production capacity, Oliver Winchester had spent considerable amounts of money to purchase new machines and increase production. Also, in the fall of 1863, Winchester had placed large parts orders that would enable him to expedite deliveries to the Army. He ordered 5,000 additional rifle levers in November 1863, and took delivery of 2,500 swivels ordered from Colt's in December. In all, the New Haven Arms Co. president did what he could to prepare for the government's purchases of Henry arms. Yet, despite his best efforts, Winchester had been greatly disappointed.[40]

## Renewed Attempts to Sell the Henry Carbine

With only the 1st D.C. Cavalry's rifle orders in hand during the early months of 1864, Oliver Winchester used his War Department contacts to arrange for further consideration of the Henry carbine. As was obvious, the best hope for selling breechloaders or repeating arms in large volume to the government was with cavalry arms.

Belatedly, Oliver Winchester was able to arrange for a further test of the Henry carbine in August 1864, at the arsenal in Washington, D.C. Lieutenant Howard Stockton and Major James G. Benton were in charge of the tests, conducted at the arsenal. Since a photograph of this arm was taken by the Army at the time of the test, the exact configuration of the Henry carbine is evident today. It appears to be the very same carbine submitted in January 1864 for trial, with perhaps some minor modifications. This time, Oliver Winchester was personally present on August 26 when the Henry carbine was tested.[41]

As reported by Lieutenant Stockton, the results were unfavorable, and even embarrassing to Oliver Winchester. Using a .44 caliber cartridge containing 36 grains of powder and a ball weighing 324 grains, accuracy tests of the Henry carbine at ranges of 350 and 500 yards were unimpressive. What proved decisive, however, was the rapidity-of-fire evaluation. The Henry carbine quickly became fouled with powder residue, and the mechanism soon worked so hard as to become "useless." The bore, with a uniform one turn in 27 inches, .075 depth of grooves, also became fouled, and the extractor frequently failed to remove the spent cartridge case.

"At this stage of the proceedings," wrote Lt. Stockton, "Mr. Winchester requested that the further trial be deferred until he should have made some alterations to prevent the fault[s]..." However, when Winchester failed to respond by October 26, 1864,

Lieutenant Stockton prepared his formal report noting that, "Since then [the August 26th test date] nothing has been heard from Mr. Winchester."[42]

Quite obviously, much of the difficulty was with the ammunition, and yet Winchester's failure to correct the half-cock safety feature the Army mentioned in its original report, which again was cited as an objection in the second trial, suggests carelessness on the part of the New Haven Arms Co. Furthermore, Winchester's inability to discover in advance the fouling difficulties suggests a further lack of testing prior to resubmission of the carbine four months after the initial failed test.

As such, it was the death knell of the New Haven Arms Company's attempts to sell the government large quantities of Henry arms. By early 1865, the Civil War was winding down, and the enormous success of less costly and more militarily practical repeating carbines, especially the Spencer, made Oliver Winchester's further efforts ineffective.

With the government's formal quest by the Laidley Board of 1865 (to evaluate all breechloading arms for adoption by both the infantry and cavalry services) the Henry rifle again failed to be approved. In the Laidley Board's final report dated April 6, 1865, of the sixty-five breechloading arms tested, the Peabody was listed as the preferred system. Yet the Spencer carbine, among magazine-fed repeaters, was cited as having "more advantages" than any other competing arm. This theme continued with the 1866 (Hancock) Board of Review, and Oliver Winchester was once again thwarted.[43]

**The Final Civil War Henry Rifle Orders**

Ironically, despite the ordnance bureaucracy's disfavor with the Henry carbine, during the organization of the United States Veteran Volunteers in the winter of 1864–1865, an opportunity to supply more standard Henry rifles to the government was suddenly at hand. Former soldiers who had not re-enlisted were to be enticed back into the Army by a substantial bounty, plus the promise of receiving a breechloading or repeating rifle when discharged (primarily Sharps were to be supplied, but Henry rifles were approved for purchase). Organized into an elite corps under the highly competent commander Major General Winfield S. Hancock, the Veteran Volunteers were intended to provide skilled, high-combat performance in bringing the Civil War to a close. Armed with the best weapons and composed of veteran soldiers, much was expected of these elite volunteers.

Yet, even with a ready, highly desirable market, the New Haven Arms Co. was again somewhat embarrassed. Due to the company's inability to sell Henry rifles or carbines to the government during the mid-war years, production had remained restricted. Although production capacity had increased beginning about December 1863, any large-quantity orders required manufacturing the arms at a sub-contract site such as Colt's, and involved the delay of at least six months. Deliveries from the existing New Haven Arms Co. factory were limited to a few hundred, not thousands of rifles per month. Thus, after the government requested price and delivery for 500 Henry rifles on March 20, 1865, two orders were processed (on April 7 and May 19, 1865, for 500 and 127 Henry rifles) but only one regiment was supplied.[44]

These last 627 U.S. martial Henry rifles were purchased to arm the 3rd Regiment U.S. Veteran Volunteer Infantry. Together with a quantity remaining on hand in U.S. arsenals from the 1st D.C. Cavalry procurements, these new Henrys were placed in the hands of Winfield S. Hancock's veterans (making a total of about 800 Henrys in the 3rd U.S.V.V.I.).[45]

## The Historic Henry Rifle — Oliver Winchester's Famous Civil War Repeater

Henry rifles in the 7000 to 8000, and 8600 to 9700 serial number ranges comprised the bulk of arms furnished by the factory for the Veteran Volunteers. As required with the 1864 government purchases, these 1865 Henrys were martially inspected and bear a cartouche or initials (such as AWM). [For a specific listing of serial numbers in Companies, B, C, I, H and K of the 3rd U.S. Veteran Volunteers, see Appendix C.] Although these new rifles were procured too late to see significant combat service in the Civil War, the 3rd Regiment's Henrys became the property of the discharged veterans when that unit was disbanded in July 1866. At this point, 681 Henry rifles went home with these discharged veteran volunteers, some of which were originally from the 1st D.C. Cavalry purchases.[46]

**Frustration Has its Beneficial Consequences**

For the New Haven Arms Co. and its Henry rifle, one of the best repeating arms to see service in the Civil War, the frustration of government disfavor in relation to its competitor arms would soon result in success. The minor faults that field use and government trials had brought into focus forced the New Haven Arms Co. to make specific improvements in the early postwar years (i.e., King's frame loading patent). With a better weapon, and not numbering among the large quantity of surplus government weapons that often flooded the market in the 1870s, the new postwar Winchester repeating arms would enjoy remarkable commercial success despite the absence of U.S. government sales.

Oliver Winchester had been both right and wrong. If disappointed in selling to the government, his scattering of a few rifles to many individuals had resulted in widespread acceptance and the broadening of his marketing base. In contrast to relying upon sales to a single large customer, which subsequently sold thousands of deactivated small arms at bargain prices when no longer needed, the firm's dispersed marketing strategy proved to be an important aspect of the Winchester firm's ultimate long-term success.

# PART III

# The Henry Rifles of Birge's Western Sharpshooters

**The 66th Illinois Infantry**

The regiment known as Birge's Western Sharpshooters was one of the most unique and effective units of the American Civil War. Organized from September to December of 1861 in various western states, the "Western Sharpshooters" as they often referred to themselves, came largely from Illinois (3 companies), Missouri (2 companies), and Ohio (2 companies). Yet many other recruits joined from neighboring states such as Indiana, Wisconsin, Minnesota and also Michigan, which raised one full company (Co. D). Only those individuals that were expert shots were recruited, and in order to qualify, a volunteer had to fire a three-shot group of 3⅓ inches maximum diameter at 200 yards — certainly not an easy task.[1]

Intended to be an elite unit used for advanced skirmish and sniper duty, Colonel John W. Birge's regiment of Western Sharpshooters was the very first such organization of its kind in the Civil War, being mustered in November 23, 1861. Col. Hiram Berdan's 1st U.S. Sharpshooters Regiment was organized about the same time in the Eastern theater, but was officially mustered in November 29, 1861.

As might be expected for such an elite unit, the response was so great that Colonel Birge's regiment was expanded to nine companies shortly after being mustered in at Benton Barracks (St. Louis, Missouri). Although the unit was designated on the rolls as the 14th Missouri Volunteer Infantry, Western Sharpshooters (often designated "W.S.S." by the men in their letters and documents) continued to be their popular name.[2]

As befitting their special status, the armament of Birge's regiment of sharpshooters was widely advertised. The Dimick "American Deer and Target Rifle," made by H.E. Dimick & Co., gunsmiths, St. Louis, Missouri, was established as the standard sharpshooters' rifle. It was a single-shot, percussion hunting or plains-type rifle with octagonal barrel, usually of half-stock, with or without a patch box, and often about .52 caliber (calibers ranged from .33 to .69). Of equal significance, this Dimick arm was designed to fire a specially configured "Swiss Chasseur"-type bullet (commonly associated with the Swiss and German Schuetzen target rifles). Accordingly, these "W.S.S." St. Louis rifles were supplied with matching bullet molds that were serially numbered to match the number stamped on each rifle's muzzle. This, of course, allowed for ready identification of the proper mold with the rifle under field conditions.

Due to the variation in bullet caliber, the handcrafted nature of these weapons, and specific military deficiencies, the logistical and technical problems with these St.

## The Historic Henry Rifle — Oliver Winchester's Famous Civil War Repeater

Louis rifles were pronounced from the very beginning. The original contract with the Dimick firm, approved September 18, 1861, was for 1,000 rifles. Yet, this quantity was not easily obtained.[3] Serial numbers on the special bullet molds have been found numbered to at least 1116, indicating that more than 1,000 rifles were eventually delivered.

Since Horace E. Dimick's firm had only about 150 rifles on hand when the first arms were procured for the Western Sharpshooters, the regiment was compelled to obtain a similarly designed rifle from various gunsmiths in the region. Consequently, many different makers' rifles were used to supply Birge's regiment. All, however, were stamped with the gun's serial number on the face of the muzzle.[4]

These civilian hunting or "St. Louis" rifles were prominently used by the Western Sharpshooters from their first encounter at Mount Zion Church, Missouri, in December 1861 to and beyond May 1863, when the regiment was serving on garrison duty at Corinth, Mississippi. As veterans of Fort Donelson, Shiloh, the siege of Corinth, and the Battle of Corinth (Oct. 1862), the regiment earned high praise for their excellent service, and at one time was even assigned to headquarters guard duty for Ulysses S. Grant.[5]

Yet, a series of unlikely events and unpromising circumstances soon befell the men of Birge's highly touted sharpshooter unit.

During November 1862, the regiment's official name was changed, the unit being redesignated the 66th Illinois Volunteer Infantry, largely due to Illinois Governor Richard Yates' influence with the War Department.[6]

In the spring of 1863, whereas other units saw active service in the Vicksburg or Middle Tennessee campaigns, the 66th Illinois (W.S.S.) remained on outpost duty near Corinth. Here the utility of their American Deer and Target rifles was questioned. As many members of the regiment had learned from their 1862 combat experience, in battle their single-shot hunting rifles were not the most utilitarian weapons. Difficult to supply with fixed or proper caliber ammunition, unable to utilize a bayonet for hand-to-hand combat, and with stocks easily damaged or broken, the St. Louis-type rifle was, at best, a sniper's arm, little adapted to the rigors of outpost and escort duty.

Company C, 66th Illinois Vol. Infantry at Camp Davies, near Corinth, Miss., ca. summer 1863. Under magnification, several Henry rifles are evident in the hands of several soldiers, including the third and twelfth from the right. (COURTESY OF BUREAU COUNTY HISTORICAL MUSEUM, PRINCETON, IL)

## PART III • The Henry Rifles of Birge's Western Sharpshooters

\* \* \*

Letter of Pvt. Marcus S. Nelson, Co. D, Birge's Western Sharpshooters, (14th Missouri Infantry; redesignated 66th Illinois Inf. In Nov. 1862), describing the fighting around Corinth, Miss., written only 14 days prior to his death in the Battle of Corinth.

*Corinth, Miss. Sept. 20th, 1862*

*Friends at home: Congratulate us! Once more there is a prospect of something being done in these parts. Price & Van Dorn are at Iuka, twenty one miles from here with a large force variously estimated at from twenty to sixty thousand men (they probably have not over twenty-five thousand) and are menacing this place. We have a heavy force at Burnsville, seven miles this side of Iuka, and it is expected that there will soon be a great battle fought at or near one of these places. The number of our troops at Burnsville is about equal to that of the enemy at Iuka, and we have the railroad to facilitate the transportation of reinforcements from this place if necessary in case of a battle. Three companies of Sharp Shooters, "D" among the number, have been out since a week ago today, skirmishing with the enemy's advance. As they left the day after I returned from the North [on furlough], I did not go with them. I should [have] though, if my feet had not been blistered with my rascally boots so that I could not march. They have had some pretty hot work out there, but at the last advice, not a man was hurt. At one time Company D was alone with the exception of three of Company F (the cowardly company), the remainder of that company having sulked, and Company K mutinied on account of the senior captain having put their [captain] under arrest for his superior [sic] bravery. The Rebels were in the edge of a piece of timber, at the top of a hill, and the S.S. were ordered by the infantry colonel who had command of the expedition to dislodge them. This of course was work for bayonets. But Co. D never flinched. Only two men in the company backed out. The rest "charged" with loud shouts up the hill in the face of the enemy's fire. It seemed like madness to rush into the woods with no arms but long range rifles [they were armed with the Dimick (St. Louis, Mo.) American Deer and Target Rifle, a sporting rifle not fitted for the bayonet], but the command was "forward" and Co. D always obeys orders. Rushing through a perfect storm of balls, they reached the top of the hill in safety, and, discharging their rifles into the woods, dashed in after the already retreating Rebels. Through this piece of woods they pursued them, and held the woods until ordered to return to Burnsville. The infantry which was ordered to support the Sharp Shooters in attempt to dislodge the Rebels from the brush followed on slowly until met by the first volley from the concealed Rebels, when they absolutely refused to proceed, and our boys were obliged to drive them out alone.*

*To show the coolness with which the boys conducted the whole thing, I will relate a couple of incidents. One of the boys, Dallas Brewster by name, when double-quicking it up the hill, saw a ball strike between his feet. He stooped down, dug it out of the dirt, put it in his pocket, and went on the same as though bullets were not flying like hail stones around him. Another dropped on one knee to load, and had just poured the powder in his gun, when an English rifle ball struck close to the toe of his boot. He picked it up, tried it in his gun, and coolly remarked, "just a fit," — "saves my going into my pouch for one," and loading his gun with the Secesh ball, he was off after the Butternuts again.*

## The Historic Henry Rifle — Oliver Winchester's Famous Civil War Repeater

*My health has been steadily improving since my return from the North, and if I can get all the milk I want, I guess I shall get along.*

*I cannot make out a meal of victuals without milk, and I have to pay fifteen cents a quart for it. When I was in Alton & St. Louis it was brought to me for five cents per quart, but here we usually have to pay twenty [cents] unless we can steal it.*

*I should like to come home and stay long enough to get in the wheat, but as we some[what] expect a "harvest" [of the enemy] here soon, I suppose Uncle Sam don't wish to spare any of his "reapers."*

*… We may be ordered away from here in the course of a few weeks, perhaps a few days. I think, however, we shall probably stay here for some time yet. Good news comes to us from Iuka tonight. The Rebels are in full retreat, and General U.S. Grant, who always does what he undertakes if he ain't drunk, is in full and <u>close</u> pursuit, bagging "game" by regiments. [Nelson refers to the Battle of Iuka, fought the previous day, Sept. 19th] A train of 21 [rail]cars has just gone out for [to bring back] prisoners, and many have been brought in before [this], within a few hours.*

*There is some prospect of taking [prisoner] the whole Rebel army. That's the way we do business in the West. We are now using every means in our power to crush the rebellion. They won't allow us to use [blacks] for soldiers, but we use them for teamsters, cooks, etc., & their women cook and wash for us, and their children wait on our officers. The most robust of them (the men) we employ in fatigue work when we have any [work] to do. They have done a "big job" of clearing for us within a few days to open a range for our siege guns to the S.W. of Corinth. I must wind up now as it is getting rather late. Write as often as you can, and believe me, as ever,*

*Your affectionate son & brother,*
*M.S. Nelson*
*W.S.S.*

*Moses Nelson/Sp'port, [Springport] Mich.*


Marcus S. Nelson, a schoolteacher from Lawrence, Van Buren County, Michigan, enlisted in Company D, Birge's Western Sharpshooters on March 10, 1862. He joined his company at Pittsburg Landing, Tennessee, on March 25, 1862, and was present at the Battle of Shiloh and the Siege of Corinth, Mississippi. Shortly after writing this letter, Private Nelson was killed in action (shot in the head) at the Battle of Corinth, Misissippi, (October 4, 1862). (COLLECTION OF WILEY SWORD)

\* \* \*

Henry rifle, serial no. 2544, within the serial range of rifles privately purchased in 1863 by the 66th Illinois Infantry, Birge's Western Sharpshooters. Rifle no. 2545 was carried by Private William P. Beesley of Co. I.
(COLLECTION OF WILEY SWORD)

## PART III • The Henry Rifles of Birge's Western Sharpshooters

From their 15-acre Camp Davies location southwest of Corinth, Mississippi, the unit was required to frequently make reconnaissance patrols into the interior. Since only a few companies of the 5th Ohio Cavalry were otherwise present at Camp Davies, the 66th's attempted scouts, patrols, ambushes and other forays into the countryside proved of limited success. Greatly responsible for this was the W.S.S.'s lack of mobility and flexibility in armament. In fact, because the Western Sharpshooters remained on foot, an effort was made to mount at least a portion of the regiment, in order to more effectively perform escort and scout duty. Of course, their muzzleloading percussion St. Louis Dimick-type rifles were entirely unsuited for mounted use. Moreover, many of these well-used hunting rifles were wearing out, or were in disrepair due to a lack of spare parts. A new rifle was clearly needed.[7]

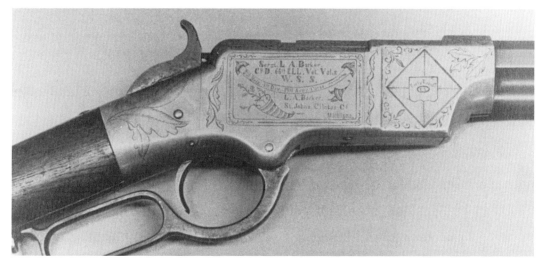

Two views of Henry rifle, serial no. 2575, Sergeant "Len" Barker's well-used rifle. In Barker's hands it saw hard service from September 1863 through the early Atlanta fighting, when Barker was wounded in the left foot at Rome Cross Roads, Georgia, on May 16, 1864. Inscribed in postwar years with the 66th Illinois' decorations, the rifle reflects the high esteem many veterans had for their "sixteen shooters." (COLLECTION OF THE MICHIGAN HISTORICAL MUSEUM, LANSING, MI)

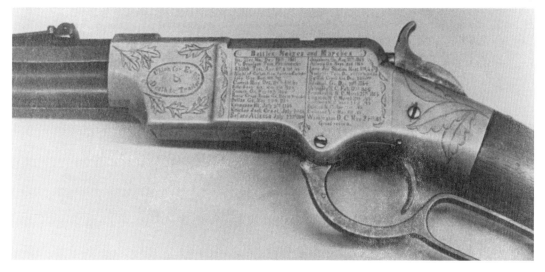

# The Historic Henry Rifle — Oliver Winchester's Famous Civil War Repeater

The answer lay in procurement of the much-talked-about "sixteen shooters," the soon-to-be-famous Henry repeating rifles that units such as Company M of the 12th Kentucky Cavalry had carried since early 1863. As might be expected, the Western Sharpshooters apparently inquired of their former arms supplier, Horace E. Dimick & Co. in St. Louis, if they could supply Henry rifles. On June 20, 1863, the Dimick firm wrote to the New Haven Arms Co. and ordered a quantity of Henry rifles. However, when the company responded on June 24, "we are not able to fill it at present on account of our [large back] orders," the Dimick firm was removed as a source of supply for Henry rifles. Accordingly, the regiment was compelled to obtain their "sixteen shooters" from a jobber such as John R. Beard at Pulaski, Tennessee, or a wholesale agent like A.B. Semple & Sons of Louisville, who in 1863 sold Henrys "by the case only."[8]

The first entry pertaining to the Henry rifle that was found in W.S.S. correspondence dates from May 25, 1863. Lieutenant William Wilson, the regiment's adjutant, noted in his journal on that day, "Some of our men are buying Henry rifles, [as] their old ones [Dimick rifles] are worthless. There was 42 [Henry rifles] that came in today [7 cases at 6 per case], costing $40 each."[9]

These, of course, were privately purchased Henrys, and resulted, in part, from the regiment's most recent pay call (April 20 when the men were paid off for four months' past service; $50 or more being received by enlisted personnel).[10]

Based on surviving documented Henry rifles associated with the Western Sharpshooters, and the chronology involving time of production and shipment to the dealer and end user (i.e., rifles made no later than April 1863), it appears many of these Henry rifles received in May 1863 were serially numbered in the 1600 and 2200 ranges. Also, due to the presence of at least one iron-framed Henry (#147) associated with a soldier in Co. G, W.S.S., some iron-framed rifles (mid-1862 production) may have been included.[11]

The success of these "sixteen shooter" armed mounted infantrymen of the 66th Illinois in skirmishes and ambuscades soon encouraged more of the regiment to buy the Henry rifle. This was especially true since "the mounted W.S.S. boys" were often designated for special escort duty. In fact, a hand-picked contingent of the 66th was selected to serve as personal escorts for district commander Brigadier General Grenville M. Dodge. All of this detail were mounted and armed with Henry rifles.[12]

The private rearmament of the Western Sharpshooters now proceeded rapidly. As recorded in various journals and personal correspondence, Henry rifles were obtained by the dozens. Many of the 66th's Michigan men (Company D) were particularly prominent purchasers of Henrys. On September 4, 1863, one Co. D soldier recorded that he had just purchased a Henry rifle for $40.00. In fact, by October 1863, so many veterans of the regiment had Henrys that on the 22nd of that month, Lorenzo Barker wrote in his journal, "the sixteen shooter boys draw horses and saddles preparing themselves to be mounted." Referring to themselves as "the sixteen shooters' cavalry," this mounted W.S.S. contingent was kept actively employed, including escorting their commander's wife to and from camp.[13]

Since much of the 66th was now being used as mounted infantry, many of the W.S.S. wanted Henry rifles fitted with slings, to prevent loss while on horseback. Some W.S.S. Henry rifles show modifications (for example, the addition of a sling ring loop to the top snap mount — see rifle #2544 on pg. 60) or exhibit specific sling wear, suggesting mounted usage.

Among various documented W.S.S. Henry rifles believed procured during this

*PART III • The Henry Rifles of Birge's Western Sharpshooters*

**Color Guard of the 7th Illinois Infantry, c.May/June 1865, with 15th Corps badges visible. They proudly display their Henrys, which helped save the day at Allatoona, Georgia, in October 1864.** (COURTESY OF THE ILLINOIS STATE HISTORICAL SOCIETY)

September/October 1863 period are rifles in the 2500 and 2600 serial number ranges, including rifle numbers 2523, 2545, 2575, 2582, 2637, 2639 and 2688.

By the end of 1863, nearly everyone in the regiment that could afford to pay for a Henry rifle had one. Based upon quarterly ordnance reports, it is evident from the Henry rifle .44 caliber ammunition on hand (furnished by the government beginning in 1864) that companies B, D, F, H and I had the majority of the 66th's Henry rifles. Company E had some, and companies A, C, G and K appear to have had only a few Henrys, considering the relative presence or absence of Henry ammunition. Interestingly, in the June 30, 1864, report (in the midst of the Atlanta campaign) a total of 180 "Dimick" rifles were reported on hand in the regiment. By the first quarter 1865 report, however, the Dimicks had entirely vanished. Only a few rifle muskets were listed in the 66th as standard government-issue arms (being used only by recruits and those unable to obtain a Henry rifle).[14]

As is evident in analyzing the periods of heavy purchase of Henrys by the regiment, their procurement appears to directly correlate with pay periods or bounty received. In November 1863, 470 men of the Western Sharpshooters re-enlisted as veterans and were due a $100 veteran bounty, plus four months back pay. They also received a 30-day veteran furlough, which began January 16, 1864. Many soldiers without Henry rifles apparently used part of this large stipend to purchase a sixteen shooter.[15]

On January 11, 1864, Sgt. Lorenzo Barker recorded in his journal: "the seventeen [sixteen] shooters or Henry rifles came to the company for those that had signed for them. The rifles [are] divided out to those in quarters." From Ordnance Department ammunition issues and other data, it is estimated that there were about 250 Henry rifles in the regiment (comprising more than one half of the unit's long arms).[16]

Thereafter, more Henrys continued to be purchased, including inter-regimental transfer, prior to the active campaigning of May 1864 (Atlanta Campaign). Even officers such as Captain William S. Boyd (who led Gen. Dodge's escort contingent) ordered their own Henry rifle, and Boyd paid $1.50 for express charges on his rifle April 4, 1864.[17]

## The Historic Henry Rifle — Oliver Winchester's Famous Civil War Repeater

For an officer or soldier about to experience active combat, the Henry's firepower represented a highly desirable asset. As witnessed by William C. Titze's purchase of Sergeant D.A. Peace's Henry rifle for $45 on February 18, 1864, many paid higher than the retail price to obtain a Henry rifle.[18]

Just how effective these Henrys were in actual combat was repeatedly demonstrated during the Atlanta Campaign. After the battle of Atlanta, Private Prosper O. Bowe of Company D breathlessly described the fighting, saying how their division had been ordered to protect the division's trains in a large open field. They had just formed in line of battle when the Rebels charged out of the woods three columns deep, shrieking an eerie yell. The 66th grimly stood its ground, reported Bowe, and not a man faltered. At the order to fire, the men cut loose with their deadly sixteen shooters, and amid the first few volleys, the column in front seemed to disappear. Although the Rebels were intent on getting to the trains and continued to fight valiantly, they fell back just as the 66th began running low on ammunition. Bowe said he fired ninety rounds without halting. The barrel of his Henry rifle was so hot that Bowe could not touch it. "[I] spit on it, and it would siz[zle]," he commented. After the fight, 700 Rebels were buried along the 66th's front, and the ground was littered with wounded, while about 2,000 prisoners were taken by their division.[19]

Earlier, at the Battle of Dallas, Georgia, May 27, 1864, Orderly Sergeant Albert C. Thompson and Private John Randall of Co. D were posted in a captured rifle pit and failed to hear the order to fall back in the face of an overwhelming attack. They "stuck to their rifle pits and worked their sixteen shooters for all they were worth," noted an eyewitness. When finally forced to surrender, they had fired their last cartridges, "and bent the barrels and broke the [stocks]" of their Henry rifles. Surprised by the fact that only two soldiers had caused so much trouble (allegedly shot ten or more men), their captors became "mad as hell," later remembered one of the prisoners. After being taken to General Pat Cleburne's headquarters for interrogation, they were sent south, and ultimately wound up at Andersonville Prison.[20]

That the Confederates highly respected and feared the Henry rifle was often demonstrated during the Atlanta fighting. After the bloody Federal repulse at Kennesaw Mountain in June, a staff officer of Cleburne's Division discovered why the enemy had managed to come "within five feet of our breastworks" even though the slaughter among their ranks was terrific: "During the armistice [arranged to keep the wounded from being burned by a ground fire that had broken out in the dry leaves] I succeeded in getting 90 rifles from the field, seven of them were Henry's patent 16 shooters."[21]

By this point in the war, the Henry rifle was a well-known weapon of awesome reputation. At Allatoona, Georgia, on October 4, 1864, the prowess of the Henry rifle was clearly demonstrated, providing a new tactical reality to ponder. The comments of a Confederate soldier from the 46th Mississippi Infantry reveal the sheer terror instilled by the deadly sixteen-shooter Yankee repeating rifles. Sergeant William P. Chambers wrote of his experience at Allatoona, saying that a senior officer of their division (that of Major General Samuel G. French) had gaily remarked that morning that taking the fortified town with its small garrison would be easy; only "a breakfast task." Yet, in a desperate frontal attack, the men of French's division came up against the 7th Illinois Infantry (same division of the 16th Corps, with the 66th Illinois), armed at their own expense with Henry repeating rifles. The Henrys put out such a volume of fire that all of the attacking Confederates were rapidly driven to cover.

## PART III • The Henry Rifles of Birge's Western Sharpshooters

While lying within thirty yards of the Yankees' breastworks, Sergeant Chambers marveled at his narrow escape. During the attack, his rifle had been struck and disabled, another bullet had penetrated his cartridge box, and he had suffered a gunshot wound in the shoulder, all within a few seconds. While crouching behind the body of a close friend, he could "hear and feel the balls as they struck the corpse." After wishfully eyeing several abandoned Henrys lying on the ground nearby, the frightened soldier concluded that matters were hopeless, and he dashed back over a rise of ground to make his escape. Facing the enemy's deadly repeating rifles was not a part of their planning, and remembering the earlier reference to this supposed easy "breakfast task," Chambers later forlornly wrote: "anybody was welcome to my share of that breakfast."[22]

The prospect of facing a superior-armed veteran Yankee soldier, wise in the ways of combat, ultimately had a profound impact on the Confederate soldier. The enemy's advantage in firepower and reduced exposure in loading changed the entire tactical battlefield. It made life a virtual hell for the Confederate infantryman who was compelled to attack rather than defend during the later stages of the war. Since the capture of a Henry rifle was of limited value (the South had no means of manufacturing metallic cartridges), it was a one-sided Northern advantage that significantly contributed to depleted Southern battlefield morale.

Conversely, to the Union soldier armed with the Henry repeating rifle, the ability of a small number of men to defeat many times their numbers instilled a spirit of confidence and inspired improved combat performance. No longer required to awkwardly load at the muzzle, a soldier armed with a Henry rifle could remain concealed or prone while reloading. Reduced combat exposure, greater firepower and the elite status that repeating rifles afforded tended to produce higher morale and also feelings of invincibility.

Yet, there seemed to be a downside to this potent combination of circumstances. The Henry repeating rifle in the hands of the deadly accurate, veteran riflemen of the Western Sharpshooters translated into a great deal of unit exposure in combat. Many senior officers were prone to heavily utilize veteran troops that were armed with repeating rifles in the most crucial combat situations, including heavy skirmish duty. During the Atlanta Campaign, the W.S.S. were under fire for 120 days, and took part in fifteen pitched battles, losing 225 officers and men killed and wounded.[23]

The extent of combat witnessed by the 66th Illinois and their deadly Henry rifles is reflected in the quantity of ammunition expended for the 2nd Division, 16th Army Corps during the Atlanta Campaign, May 4th to Sept. 8th 1864.

| | |
|---|---|
| Elongated ball .58 caliber | 714,150 (88.3%) |
| Henry rifle .44 caliber | 83,500 (10.3%) |
| Spencer rifle .52 caliber | 11,088 (1.4%) |

Since the 3rd Brigade was detached at Rome, Georgia, only the 1st and 2nd brigades saw heavy fighting during the July 1864 combat around Atlanta. Among these two brigades was included only one unit heavily armed with the Henry rifle, which was the 66th Illinois. The 7th Illinois, which was also heavily Henry armed, was with the 2nd Division's 3rd Brigade at Rome and saw very limited action in the campaign.

The significance of this expenditure is evident when comparing the Henry rifle rounds fired versus that for other infantry weapons during the entire Atlanta Campaign.

## The Historic Henry Rifle — Oliver Winchester's Famous Civil War Repeater

|  | Expended Ammunition total rifle rounds | Henry rounds |
|---|---|---|
| Army of the Cumberland | 11,815,308 | 10,249 (.0001%) |
| Army of the Tennessee | 8,182,645 | 93,655 (1.2%) |
| Army of the Ohio | 1,875,559 | 23,300 (1.2%) |
| **Total** | 21,873,512 | 127,204 (.0058%) |

The 66th's expenditure of 83,500 rounds (89.2%) of the 93,655 Henry cartridges fired by the Army of the Tennessee demonstrates the prolific service of that regiment during the Atlanta fighting. The 66th Illinois fired nearly two thirds (65.6%–83,500/127,204) of the Henry rifle ammunition expended during the entire Atlanta Campaign.

Again under fire during actions at Savannah, Georgia, and in the Carolinas, the 66th Illinois acquitted itself with high honors, particularly on March 21, 1865, in heavy fighting near Bentonville, N.C.

Having participated in Sherman's March through Georgia, and the Carolinas Campaign, the regiment marched in the Grand Review at Washington, D.C. on May 24, 1865, and was finally mustered out at Louisville, Kentucky, July 7, 1865.[25]

Of course, nearly all W.S.S. veterans took their Henry rifles home with them, they being privately owned arms. Some of these weapons subsequently became practical tools of the hunt, or were used in frontier service by those who went west. Otherwise, many were kept or displayed as trophies of the late war, serving as mementos for a generation of veterans who remembered with pride their sixteen shooters' important role.

Just how revered these Henrys remained to the men was demonstrated long after the war. During the first annual reunion of Co. D, W.S.S. in 1884, their historian recorded: "Ren Barker was accompanied [to the reunion] by his old friend in war and companion in peace, the Henry rifle carried by him through the war. The old "bull dog" bore the names of the battles, skirmishes and marches of the company, neatly engraved on its brass mountings [side plates]. It was handled carefully and reverently by all, and is prized by the owner above money."[26] Barker's Henry rifle, serial no. 2575, is now in the Michigan Historical Museum, Lansing, Michigan. As is reflected by this weapon, it was a popular practice to inscribe and/or engrave these brass-framed weapons after the war to perpetuate their status as an honored keepsake.

Indeed, contrary to what many collectors might assume today, very few Henry rifles were inscribed or engraved contemporary to their Civil War period of use. The vast majority of "military inscribed" Henrys observed today were so marked long after their period of active service. Indeed, it became sort of a fad for the aged veterans of the 66th Illinois (Western Sharpshooters) and others to decorate or inscribe their Henrys in the postwar era. If only to carry these esteemed rifles to reunions and show them off as mementos of the war, and to help retell vivid stories, many W.S.S. Henry rifles again saw "reunion" service in the hands of their proud owners. Of course, only a few of the regiment's original Henrys were so marked, and the majority of sixteen shooters of the W.S.S. today can only be approximated by serial number and wear patterns.

The importance of the Western Sharpshooters, originally armed with their Dimick St. Louis plains rifles, and later with Henry repeating rifles, can only be surmised today through the memoirs and correspondence of the soldiers. That their Henry rifles

## PART III • The Henry Rifles of Birge's Western Sharpshooters

performed arduous service and contributed significantly to the Union campaign successes of 1864 and 1865 is certain.

In all, the 66th Illinois' Henry rifles contributed an important legacy for historians and collectors. Yet, their rarity as collector's weapons has often been overlooked. If survival rates of similarly hard-used historical arms are considered, it is estimated that of the Western Sharpshooters' approximately 250 Henry rifles, less than 50 survive today.

### Serial Numbers of Henry Rifles Identified with the 66th Illinois Infantry (Birge's Western Sharpshooters)

#147    iron frame, identified to J. Marshall Co. G [reference: Ken Baumann, *Arming Those Suckers*, p. 153]

#213    brass frame, "J.W. Schuessler, Co. F, 66th Ill./Birge's Sharpshooters, July 1863" [ref. Howard M. Madaus, "The First 1,500 Henry Rifles"]

#1398    "Louis Quinius/Co. B, W.S.S." [ref. R.L. Wilson, *Winchester: An American Legend*, pp. 16–17, collection of C.W. Slagle]

#1606    "Milo N. Damon/Co. A, W.S.S." [reference: Doug Bennick, Orange, CA]

#1614    documented to George W. Yarington, Co. D (Don Troiani)

#1639    "Asahel Horton/Co. B, W.S.S." (reference: John E. Parsons, *The First Winchester*, 1969 edition [appendix xiv])

#1672    "Enos W. Tyler/Co. D, W.S.S." (reference: Doug Bennick, Orange, CA)

#1692    "C. Parish, Co. F, 66th Ill., 2d Brig. 4th Div." (ref.: Ken Baumann, *Arming Those Suckers*, p. 153)

#2274    Pvt. Hiram S. Vinson, Co. K; from 1864 diary entry; enlisted 3-5-62, must. out 1865 (collec. of Steve Altic, Cleveland, Ohio, per Dan Fagen, Florissant, Mo.)

#2287    "David Padgett" "Co. K/66th Ill. W.S.S." "L/341" (ref.: Richard Ellis Auction Catalog 10-17-1993, lot #34)

#2523    "J.W. VanBrocklin/Co. D, 66th Ill. Western Sharpshooters" (reference: George Madis, *The Winchester Book*, p. 48)

#2545    Private William P. Beesley, Co. I — from his 1864 pocket diary entry "number of my Henrey [sic] Rifle 2545"; (reference: Dan Fagen, Florissant, Mo.)

#2575    "Sergt. L.A. Barker/Co. D, 66th Ill. Vet. Vols. W.S.S." Mich. Historical Museum Coll. (reference: Ken Baumann, *Arming Those Suckers*, p. 112 ff.)

#2582    "L.P. Tallman/Co. A/Western Sharp Shooters/Vet. Vols." (reference: John E. Parsons, *The First Winchester*, 1969 ed. [appendix xx])

#2637    Private David Litherland, Co. I — died 4-10-64, from diary entry; (reference: Dan Fagen, Florissant, Mo.)

#2639    "Chas. Webster/Co. D, W.S.S. Vet. Vols." (reference: Ken Baumann, *Arming Those Suckers*, p. 153)

#2688    "Thomas E. Gleason. Co. D, W.S.S." (ref.: Jackson Arms Catalog #20 — Colln. Glenn Hanna, Battle Creek, Mich.)

#2900    "J.M. Bodge. Co. H. W.S.S." (ref.: Cliff Eckle, curator, colln. of the Historical Society, Columbus, Ohio)

# The Historic Henry Rifle — Oliver Winchester's Famous Civil War Repeater

Based upon 470 men re-enlisting for veteran service in January 1864, an estimated 250 Henry rifles were in use by the 66th Illinois (W.S.S.) in January 1864. Per 1st Quarter 1864 Ordnance report — 220 other longarms (Dimicks & rifle muskets) were reported issued to the 10 companies. According to the data on ammunition below, it is estimated that in the 66th Illinois there were: Many Henry rifles in Co's B, D, F, H, I (5 companies); some Henry rifles in Co's E (one company); a few Henry rifles in Co's A, C, G, K (4 companies).

## Henry Rifle Ammunition Reported in the 66th Illinois
*Source: U.S. Ordnance Records, RG 156, National Archives Summary Statements for Small Arms in the Hands of Troops.*

Beginning in 1864, Henry rifle ammunition was provided by the federal government to units armed with these rifles, even though the arms were privately purchased. By checking quarterly ordnance summary statements following this date, Henry rifle presence is reflected in the government ammunition provided and on hand. However, since the 66th's Henry rifles were not U.S. government issued, they are not listed on these ordnance reports, which show *only U.S. government property.*

Please note that Company D is among the 66th's most heavily armed with Henry rifles from the earliest purchases. Even though for the third quarter ending Sept. 30, 1863, no Henry U.S.-issued metallic cartridges are reported on hand, about a dozen Henrys were then estimated to be in use within that unit. Co. D also had 46 Dimick rifles in inventory. During the fourth quarter ending Dec. 31, 1863, Co. D reported 42 Dimick rifles on hand. Since the government began supplying Henry rifle ammunition to the 66th Illinois during the 1st Quarter of 1864, prior reports would not list privately purchased Henry rifle ammunition on hand.

### First Quarter ending March 31, 1864
**307 total arms (81 Dimick rifles) "HY" [Henry] ammunition on hand**

|        | HY cartridges | other rifles |
|--------|---------------|--------------|
| Co. A  | x             | 15 rifle muskets .58 cal. |
| Co. B  | 1,250         | 22 Dimick rifles. |
| Co. C  | x             | 41 rifle muskets .58 cal. |
| Co. D  | 11,016        | 20 rifle muskets .58 cal. |
| Co. E  | 2,000         | 18 rifle muskets .58 cal. |
| Co. F  | 1,100         | 20 rifle muskets .58 cal. + 22 Dimick |
| Co. G  | 507           | 17 rifle muskets .58 cal. |
| Co. H  | 2,678         | 47 rifle muskets .58 cal. |
| Co. I  | 2,560         | 20 rifle muskets .58 cal. + 37 Dimick |
| Co. K  | no report     | |

## PART III • The Henry Rifles of Birge's Western Sharpshooters

**Second Quarter ending June 30, 1864 "HY" [Henry] ammunition on hand**

|       | HY cartridges | other rifles |
|-------|---------------|--------------|
| Co. A | x             | 14 rifle muskets |
| Co. B | 1,500         | 1 rifle muskets + 20 Dimick rifles |
| Co. C | x             | 36 Dimick rifles |
| Co. D | no report     |  |
| Co. E | 400           | 18 rifle muskets + 48 Dimick rifles |
| Co. F | no report     |  |
| Co. G | x             | 22 rifle muskets + 12 Dimick rifles |
| Co. H | 1,078         | 2 rifle muskets + 29 Dimick rifles |
| Co. I | 2,000         | 27 rifle muskets + 7 Dimick rifles |
| Co. K | x             | 28 Dimick rifles |
| Detail 805 | | |

**Third Quarter ending Sept. 30, 1864 "HY" [Henry] ammunition on hand**

|       | HY cartridges |
|-------|---------------|
| Co. A | 10            |
| Co. B | 1,350         |
| Co. C | x             |
| Co. D | 230           |
| Co. E | x             |
| Co. F | 1,000         |
| Co. G | no report     |
| Co. H | 1,200         |
| Co. I | 2,000         |
| Co. K | 500           |

**Fourth Quarter ending Dec. 31, 1864 "HY" [Henry] ammunition on hand**

|        | HY cartridges |
|--------|---------------|
| Co. A  | 610           |
| Co. B  | 600           |
| Co. C  | x             |
| Co. D  | 650           |
| Co. E  | x             |
| Co. F  | x             |
| Co. G  | 100           |
| Co. H  | 1,300         |
| Co. I  | 2,100         |
| Co. K  | 390           |
| stores | 32,192        |

**First Quarter ending Mar. 31, 1865 "HY" [Henry] ammunition on hand**

|        | HY cartridges |
|--------|---------------|
| Co. A  | x             |
| Co. B  | 200           |
| Co. C  | 1,200         |
| Co. D  | 750           |
| Co. E  | 720           |
| Co. F  | 690           |
| Co. G  | 560           |
| Co. H  | 650           |
| Co. I  | 1,220         |
| Co. K  | 330           |
| stores | 16,900        |

**General Summary**

Reflecting on the popularity of Henry rifles in the 66th Illinois, three major procurements for the regiment by private purchase are noted:

1. Forty-two rifles obtained by members of various companies while at Corinth, Miss. in May 1863; estimated serial range 1650 to 2300 (based on documented rifles) with perhaps a few iron-frame rifles (e.g., #147) included therein;
2. Many additional Henrys were obtained while at Corinth, Miss., especially by

## The Historic Henry Rifle — Oliver Winchester's Famous Civil War Repeater

Co. D during Sept. 1863 — est. SN# range 2520–2700 (based on documented rifles);
3. Purchase of Henry rifles by various members of the regiment at about $40 each while at Camp Davis, Tenn., and received for distribution about Jan. 11, 1864. Estimated serial range 3900+ and 4000–4500. This assumes the block U.S. martial serial no. range of 3000 to 4000 was set aside in late 1863, which may have resulted in out-of-sequence serial numbers being shipped from the factory; i.e., rifles in the low 4000 serial range leaving the factory prior to the 800 rifle U.S. martial shipments (serial no. range of 3000–4000) of March 1864. The basis for this late purchase by the 66th Illinois' veteran soldiers is believed to be the bounty for reenlistment of $100 paid to each enlisted man after Nov. 1863, which, plus back pay due, provided ample funds for buying the very expensive Henry rifle ($42). Further, as is documented in various journal and diary entries, private soldiers sold or exchanged their Henrys when leaving the regiment (e.g., Wm. C. Titze's purchase of Sgt. D.A. Peace's Henry on Feb. 18, 1864, for $45).

**Birge's Western Sharpshooters:**
- Organized Sept./Oct. 1861
- Designated 14th Missouri Infantry Nov. 23, 1861
- Designated 66th Illinois Infantry Nov. 20, 1862
- Veterans furloughed Jan. 16 to March 8, 1864
- Mustered out July 7, 1865.

**Major Combat —**
- Mount Zion, Missouri, December 28, 1861
- Fort Donelson, Tennessee, February 13–15, 1862
- Shiloh, Tennessee, April 6–7, 1862
- Siege of Corinth, Mississippi, April–May 1862
- Iuka, Mississippi, September 19, 1862
- Corinth, Mississippi, October 3–4, 1862
- Hatchie Bottoms, Mississippi, November 28, 1862
- Guerrilla fighting in the vicinity of Corinth, Mississippi, 1863
- Atlanta, Georgia Campaign 1864, including Resaca, Dallas, Lone Mountain, New Hope Church, Big Shanty, Kennesaw Mountain, Marietta, Ezra Church, Battle of Atlanta, Jonesboro, and Lovejoy's Station
- Sherman's March to the Sea, 1864
- Savannah, Georgia, December 31, 1864
- Columbia, S.C., February 17, 1865
- Camden, S.C., February 20, 1865
- Bentonville, N.C., March 21, 1865
- Kingston, N.C., March 24, 1865
- Goldsboro, N.C., March 26, 1865
- Raleigh, N.C., April 12, 1865
- Richmond, Virginia, May 13, 1865.

# Appendix A

## Estimated Henry Production by Serial Numbers

This estimate is based largely upon notes and factory data recorded in the New Haven Arms Co. Letter Book of letters sent, pages 97–500, Oct. 8, 1862 to Dec. 12, 1863, Cody Firearms Museum, Buffalo Bill Historical Center, Cody, Wyoming. See also Williamson, *Winchester The Gun That Won the West*; and Parsons, *The First Winchester*. There are no known surviving serial number and delivery records remaining from the New Haven Arms Co., although these journals were kept and are mentioned in company correspondence. Also, please be aware that the Henry rifle was manufactured in production lots using various contract workers within the plant. Accordingly, it is certain that *rifles did not leave the factory in precise serial number sequence.* However, since the demand for Henry rifles was generally greater than the supply during the course of the War, it is estimated that few rifles remained on hand after being finished. Shipments of rifles may be considered to approximate their chronological production. This suggests that shipments roughly corresponded to their proper serial number sequence. The one notable exception to this would involve deluxe, engraved rifles, which required additional time to prepare. Their shipment would generally be of later date by serial number than the corresponding production rifle. (For example, SN#1978, was manufactured c.April 1863, and then personally presented by Oliver Winchester to a Connecticut captain, Oct. 23, 1863.)

**NOTE:** The following list reflects the inclusion of iron-frame Henry rifles within the same common serial number sequence (see Appendix E — Iron-Framed Henry Rifles).

| DATE | PRODUCTION | SER.# AT END OF MO. |
|---|---|---|
| Apr.–May–June 1862 | 300 | 300 |
| July 1862 | 125 | 425 |
| Aug. 1862 | 125 | 550 |
| Sept. 1862 | 150 | 700 |
| Oct. 1862 | 200 | 900 |
| Nov. 1862 | 200 | 1100 |
| Dec. 1862 | 200 | 1300 |

**Jan. 1, 1863** (1,300 rifles produced in 1862; inclusive of approx. 200 iron-frame rifles)

| DATE | PRODUCTION | SER.# AT END OF MO. |
|---|---|---|
| Jan. 1863 | 200 | 1500 |
| Feb. 1863 | 200 | 1700 |
| Mar. 1863 | 200 | 1900 |
| Apr. 1863 | 200 | 2100 |
| May 1863 | 200 | 2300 |
| June 1863 | 200 | 2500 |

order from U.S. Gov't. June 20, 1863 — 241 rifles filled 7-23-63

| DATE | PRODUCTION | SER.# AT END OF MO. |
|---|---|---|
| July 1863 | 200 | 2700 |
| Aug. 1863 | 200 | 2900 |
| Sept. 1863 | 200 | 3100* |
| Oct. 1863 | 200 | 3300* |

order from U.S. Gov't. open market — 60 rifles filled 10-31-63

| DATE | PRODUCTION | SER.# AT END OF MO. |
|---|---|---|
| Nov. 1863 | 225 | 3525* |
| Dec. 1863 | 250 | 3775* |

## The Historic Henry Rifle — Oliver Winchester's Famous Civil War Repeater

*NOTE: Out-of-serial-number-sequence shipment of Henry rifles (serial numbers in 4000 range) may have occurred during this period due to the assignment of the 3000–4000 block serial numbers to the U.S. Government for the 1st D.C. Cavalry's additional companies.

| DATE | PRODUCTION | SER. # AT END OF MO. |
|---|---|---|
| **Jan. 1, 1864** (2,475 rifles produced in 1863) | | |
| order from U.S. Gov't. 12-31-1863 — 800 rifles filled 3-9-64 | | |
| Jan. 1864 | 250 | 4025 |
| Feb. 1864 | 275 | 4300 |
| Mar. 1864 | 275 | 4575 |
| Apr. 1864 | 300 | 4875 |
| May 1864 | 300 | 5175 |
| Jun. 1864 | 300 | 5475 |
| Jul. 1864 | 300 | 5775 |
| Aug. 1864 | 300 | 6075 |
| Sep. 1864 | 325 | 6400 |
| Oct. 1864 | 325 | 6725 |
| Nov. 1864 | 350 | 7075 |
| Dec. 1864 | 350 | 7425 |
| **Jan. 1, 1865** (3,550 rifles produced in 1864) | | |
| Jan. 1865 | 350 | 7775 |
| Feb. 1865 | 350 | 8125 |
| Mar. 1865 | 375 | 8500 |
| Apr. 1865 | 425 | 8925 |
| order from U.S. Gov't. Apr. 7, 1865 — 500 rifles filled 4-19-65 | | |
| May 1865 | 475 | 9400 |
| order from U.S. Gov't. May 16, 1865 — 127 rifles filled 5-23-65 | | |
| June 1865 | 400 | 9800 |
| July 1865 | 400 | 10200 |
| Aug. 1865 | 200 | 10400 |
| Sept. 1865 | 150 | 10550 |
| Oct. 1865 | 150 | 10700 |
| Nov. 1865 | 100 | 10800 |
| Dec. 1865 | 100 | 10900 |

**Jan. 1, 1866** (3,475 rifles produced in 1865)

Jan. 1866 to end of regular Henry production in the summer of 1867 following retooling and changeover: (approximately 110 per month avg.) about 2,000 total regular production to about s.n. 12850

The highest serial number for a Henry rifle in the "traditional" range is 12832, yet several examples of standard Henry rifles have been reported with higher serial numbers, including into the 14000 range (no. 14262), see R.L. Wilson, *Winchester...*, p. 17, also *Man at Arms* Magazine, Jan./Feb. 1992, pp. 8 ff., "Winchester Model 1866 Serial Numbers — Another Perspective," by Wiley Sword.

# APPENDIX B

## U.S. Government Purchases of the Henry Rifle 1862–1865

| DATE | SOURCE | QTY. | EST. SER. NOS. |
|---|---|---|---|
| 4-9-1863 | Merwin & Bray, NY may have involved the Lafayette C. Baker requisition | 1 | unknown |
| 6-19-1863 (shipped) | John W. Brown, agent to supplement factory's shortage on the 1st D.C. Cavalry formal order for 120 | 50 | 1300–2100 range |
| 6-20-1863 (shipped) | New Haven Arms Co. per formal War Dept. order for 120, reduced from Baker's reqn. for 240, for 1st D.C. Cav. | 80 | 2100–2300 range |
| 7-21-1863 (shipped) | New Haven Arms Co. balance of the "Baker" requisitioned quantity of 240; order for 120 is amended | 110 | 2300–2600 range |
| 9-19-1863 | New Haven Arms Co. for trial or evaluation | 1 | unknown |
| 10-26-1863 (order date) | John W. Brown, agent per requisition of L.C. Baker for Co. D, 1st D.C. Cavalry, factory unable to fill | 60 | 1300–3000 range |
| 12-30-1863 (ordered) | New Haven Arms Co. | 800 | 3000–4000 (bear "C.G.C." insp. mks.) |
| 3-9-1864 (shipped) | for additional companies of the 1st D.C. Cavalry | | |
| 6-17-1864 | New Haven Arms Co. for trial or evaluation | 1 | unknown |
| 4-7-1865 | New Haven Arms Co. (ordered) | 500 | 7000–8000 range + a few in 6800s |

## The Historic Henry Rifle — Oliver Winchester's Famous Civil War Repeater

| DATE | SOURCE | QTY. | EST. SER. NOS. |
|---|---|---|---|
| 4-19-1865 (shipped) | for the 3rd U.S. Veteran Vol. Infantry | | |
| 5-16-1865 (ordered) | New Haven Arms Co. | 127 | 8400–9400+ range |
| 5-23-1865 (shipped) | for the 3rd U.S. Veteran Vol. Infantry | | |
| 11-7-1865 | New Haven Arms Co. | 1 carbine @ $35 | |
| | unknown, believed to be the special Henry test carbine prepared in Jan. 1864 and later submitted for government tests in 1864 and 1865 | | |

**TOTAL PURCHASES – 1,731**

The U.S. Government purchased about 18.4% of Henry rifle factory production through May 1865 (1,731/9,400). The highest known U.S. martial serial number is 9701, which may be a replacement arm.

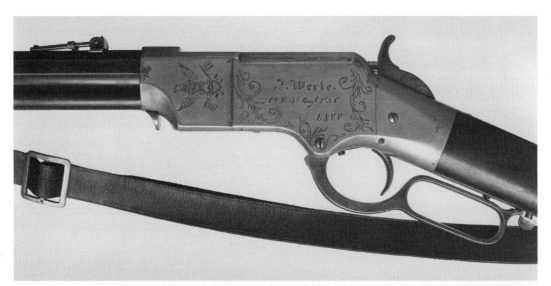

Henry rifle, serial number 7582, manufactured in late 1864, is complete with an original Henry rifle sling. It was originally owned by Pvt. Jacob Werle who served in two Pennsylvania infantry regiments (the 21st and 98th) before enlisting in the 3rd U.S. Veteran Volunteer Infantry in March of 1865. (THOMAS M. HARDMAN COLLECTION)

# APPENDIX C

## Serial Numbers of Henry Rifles
## Reported by the 3rd Regiment U.S. Vet. Vol. — 1865
## Companies B, C, I, H, K

| | | | | | | | | | | |
|---|---|---|---|---|---|---|---|---|---|---|
| 1392 | 2055 | 3035 | 3134 | 3219 | 3330 | 3418 | 3551 | 3613 | 3715 | 3805 |
| 1491 | 2057 | 3038 | 3149 | 3220 | 3332 | 3419 | 3557 | 3632 | 3735 | 3807 |
| 1492 | 2076 | 3039 | 3164 | 3296 | 3344 | 3423 | 3570 | 3652 | 3746 | 3809 |
| 1592 | 2101 | 3055 | 3184 | | 3354 | 3433 | | 3655 | 3751 | 3810 |
| 1732 | 2312 | 3060 | | | | 3440 | | 3666 | 3759 | 3838 |
| 1798 | 2314 | 3062 | | | | 3462 | | | 3775 | 3852 |
| 1804 | 2547 | 3079 | | | | 3468 | | | 3777 | 3857 |
| 1898 | 2562 | 3083 | | | | 3469 | | | 3782 | 3878 |
| 1935 | 2572 | 3098 | | | | 3485 | | | 3788 | 3879 |
| 1939 | 2573 | | | | | 3486 | | | 3799 | 3881 |
| 1954 | | | | | | 3498 | | | | 3884 |
| | | | | | | | | | | 3885 |
| | 2826 | | | | | | | | | 3893 |

---

| | | | | | | | | | | |
|---|---|---|---|---|---|---|---|---|---|---|
| 3901 | 6806 | 7100 | 7200 | 7315 | 7405 | 7500 | 7603 | 7700 | 7801 | 7900 |
| 3902 | 6809 | 7101 | 7203 | 7320 | 7408 | 7503 | 7609 | 7706 | 7807 | 7954 |
| 3917 | | 7102 | 7204 | 7334 | 7409 | 7505 | 7610 | 7734 | 7812 | 7967 |
| 3918 | | 7108 | 7209 | 7347 | 7412 | 7506 | 7615 | 7755 | 7822 | |
| 3919 | 7074 | 7112 | 7210 | 7355 | 7415 | 7508 | 7619 | 7774 | 7830 | |
| 3924 | 7075 | 7114 | 7214 | 7365 | 7416 | 7509 | 7642 | | 7844 | |
| 3935 | 7097 | 7117 | 7217 | 7374 | 7418 | 7511 | 7646 | | 7854 | |
| 3938 | | 7118 | 7227 | 7384 | 7419 | 7520 | 7651 | | 7866 | |
| 3941 | | 7124 | 7235 | 7395 | 7426 | 7528 | 7653 | | 7878 | |
| 3943 | | 7127 | 7239 | | 7437 | 7532 | 7663 | | 7881 | |
| 3944 | | 7130 | 7245 | | 7439 | 7533 | 7664 | | 7882 | |
| 3949 | | 7134 | 7250 | | 7453 | 7539 | 7665 | | 7889 | |
| 3951 | | 7135 | 7259 | | 7456 | 7541 | 7672 | | 7898 | |
| 3954 | | 7136 | 7261 | | 7458 | 7551 | 7692 | | | |
| 3955 | | 7139 | 7266 | | 7465 | 7553 | 7694 | | | |
| 3956 | | 7141 | 7277 | | 7466 | 7565 | 7698 | | | |
| | | 7146 | 7278 | | 7468 | 7567 | | | | |
| | | 7147 | 7283 | | 7470 | 7574 | | | | |
| | | 7149 | 7292 | | 7471 | 7577 | | | | |
| | | 7150 | 7297 | | 7479 | 7582 | | | | |
| | | 7172 | 7299 | | 7482 | 7583 | | | | |
| | | 7173 | | | 7485 | 7588 | | | | |
| | | 7175 | | | 7490 | 7591 | | | | |
| | | 7180 | | | 7493 | 7592 | | | | |
| | | 7189 | | | 7495 | 7594 | | | | |
| | | 7190 | | | | | | | | |
| | | 7195 | | | | | | | | |
| | | 7198 | | | | | | | | |

## APPENDIX C  (continued)

| | | | | | | | | | |
|---|---|---|---|---|---|---|---|---|---|
| 8480 | 8600 | 8702 | 8805 | 8949 | 9105 | 9202 | 9300 | 9483 | 9701 |
| | 8605 | 8708 | 8810 | | 9116 | 9207 | 9310 | | |
| | 8610 | 8710 | 8816 | | 9131 | 9213 | 9316 | | |
| 8577 | 8617 | 8713 | 8818 | | 9137 | 9225 | 9323 | | |
| 8592 | 8622 | 8725 | 8820 | | 9168 | 9226 | 9340 | | |
| | 8623 | 8726 | 8853 | | 9169 | 9230 | 9346 | | |
| | 8627 | 8728 | 8863 | | 9174 | 9250 | 9362 | | |
| | 8634 | 8735 | 8868 | | 9193 | 9251 | 9369 | | |
| | 8635 | 8747 | 8877 | | | 9257 | | | |
| | 8639 | 8751 | 8878 | | | 9264 | | | |
| | 8648 | 8752 | 8885 | | | 9270 | | | |
| | 8650 | 8755 | 8891 | | | 9279 | | | |
| | 8652 | 8757 | 8895 | | | 9284 | | | |
| | 8667 | 8758 | | | | 9287 | | | |
| | 8673 | 8761 | | | | 9287 | | | |
| | 8680 | 8763 | | | | 9292 | | | |
| | 8696 | 8764 | | | | 9293 | | | |
| | | 8766 | | | | 9299 | | | |
| | | 8768 | | | | | | | |
| | | 8770 | | | | | | | |
| | | 8773 | | | | | | | |
| | | 8774 | | | | | | | |
| | | 8775 | | | | | | | |
| | | 8778 | | | | | | | |
| | | 8787 | | | | | | | |
| | | 8791 | | | | | | | |
| | | 8797 | | | | | | | |
| | | 8798 | | | | | | | |

It is estimated that in the above list, rifles no. 1392 through 2573 represent arms purchased for the 1st D.C. Cavalry (battalion) in 1863, and were supplied by John W. Brown, agent, Columbus, Ohio, as well as the New Haven Arms Co. Rifles no. 2826 through 3956 are believed to be the March 1864 supplemental purchases supplied by the factory for the additional 1st D.C. Cavalry companies. Later, these rifles, 1392 through 3956, are believed to have been turned in (mostly by the 1st D.C. Cavalry), or remained unissued, and were in the arsenal when needed for the 3rd U.S. Veteran Volunteers in 1865. Since only 800 Henrys were obtained by the U.S. Government in 1864 from the 1,000 block (nos. 3000–4000), some U.S.-inspected rifles may have remained on hand at the factory until 1865. This is suggested by the prevalence of rifles in the 3800 and 3900 ranges listed above for the 3rd U.S.V.V.

Summary:
|  |  |  |  |  |
|---|---|---|---|---|
| | 21 | Henrys 1392 thru 2573 | | |
| | <u>79</u> | <u>Henrys 2826 thru 3956</u> | 100 Total | **Combined** |
| | 150 | Henrys 6806 thru 7967 | | **348 TOTAL** |
| | <u>98</u> | <u>Henrys 8480 thru 9701</u> | 248 Total | |

*Appendix C*

Other Henry rifles reported with U.S. Martial associations:

        2797  Co. F, 97th Indiana
        3117  Co. K, 7th Illinois Vols. 12-1864
        3233  Co. F, 97th Indiana
        3347      "
        3509      "
        3532      "
        4228  Co. I, 73rd Illinois Vols. 11-1864
        4392  Co. E, 36th Illinois Vols. 2-1865

Henry rifle serial no. 8622 inscribed:

                "T. Pfeifer, Co. B"
            [3d Regt. U.S. Vet. Vol. Inf.]

Reference: Frank Mallory: Springfield Research Service, issue no. 78 of *U.S. Martial Arms Collector*, October 1996, pp. 78–11, 12

    Documentation of rifles bearing the above serial numbers may be obtained from Frank Mallory, Springfield Research Service, P.O. Box 4181, Sliver Spring, MD 20904, (301) 622-4103.

# APPENDIX D

## Partial List of Historically Identified Henry Rifles

The following list of inscribed or otherwise historically documented Henry rifles is by no means complete or likely to be totally accurate. These are arms that have been reported as so inscribed or identified. There is no guarantee that any inscription is original, or that the serial numbers reported are accurate. This list is presented merely as a compilation from various sources, including dealer lists, publications, personal observation and collector records.

| No. | Identification | Remarks |
|---|---|---|
| 1 | Edwin M. Stanton/Secretary of War (plaque) | |
| 6 | Lincoln/President/U.S.A. (plaque) | brass, engr. |
| 7 | W.T. Moyers, Atlanta, Ga. | brass, engr. |
| 9 | Gideon Wells-Sec. Navy (plaque) | brass, engr. |
| 14 | S. Hodgson | |
| 19 | allegedly presented to Geo. D. Prentice, Louisville, Ky. | brass, engr.? |
| 147 | J. Marshall Co. G, 66th Illinois [listed] | iron frame |
| 165 | 5th Tenn. Cav. July 27, 1862 (crudely inscribed - reported) | iron frame |
| 213 | J.W. Schuessler, Co. F, 66th Ill. Birge's Sharpshooters, July 1863 | brass |
| 287 | Wm. S. Skelton (inscribed) [Lieut. Co. E, 1st Ark. Cav. K.I.A. 10-1862] | brass |
| 325 | Col. N.P. Chipman/Maj. Gen. Curtis staff/pres. by Maj. A.C. Ellithorpe/1st Indiana Regt. (inscribed) [relic - dug] | brass buckhorn sight |
| 324, 359, 391, 706 | Returned to the factory for repair by John W. Brown, agent, Columbus, Ohio in October 1862, returned Oct. 20th to Brown | |
| 467 | Robert S. Denee, Ontario (inscribed) | brass |
| 504 | F.W. Binger, Salt Creek, Neb. 1864 (crude inscription) | brass |
| 562 | "L.L." [Louisville Legion] also 5th Ky. Inf. U.S.V." (inscribed) | brass |
| 698 | Capt. H. Bingham (inscribed) | |
| 771 | Capt. J.A. Smith/Co. E/7th Ill. Inf. (inscribed – but not of period) | |

| No. | Identification | Remarks |
|---|---|---|
| 976 | A.W. Morris/Co. D/16th Ill. Vet. Vol. Inf. (postwar inscribed) | |
| 1006 | Maj. T.J. Swan/Surg. 12th Ky. Cav. presened by Capt. Jas. Willson (inscribed) | |
| 1008 | Sergt. Jackson – W.J. Jackson 65th Ind. Vol. Co. A (inscribed) | tacks/capt. from Indians |
| 1116 | Daniel McCook - paymaster U.S. Vols. KIA 7-18-63 [identified] | |
| 1187 | H.C. Stout, 65th Ind. Vol. Inf. (crude inscription) | worn condition |
| 1225 | A.C. Osborn (inscribed on carrier) | |
| 1228 | Pres. by Wells Fargo & Co. to Stephen Venard for his gallant conduct May 16, 1866 (plaque) | |
| 1302 | Maj. G.L. Febiger to T.R Biggs, Cin. O (inscribed) | |
| 1348 | Capt. A.L. Fahnestock (inscribed) [86th Ill. Inf.] | photo exists of Capt. F. holding |
| 1398 | Louis Quinius/Co. B, W.S.S. (inscribed) [66th Ill. Inf.] | |
| 1434 | W.F. Lunt/1st D.C. Cav. Co. (crude inscription) | buckhorn sight |
| 1534 | S. Wright, K$^W$, March 29th 1863 (inscribed) | |
| 1606 | Milo N. Damon/Co. A, W.S.S. (inscribed) [66th Ill. Inf.] | |
| 1614 | George W. Yerington, Co. D, 66th Ill. (documented) | |
| 1639 | Asahel Horton/Co. B W.S.S. (inscribed) [66th Ill. Inf.] | |
| 1672 | Enos W. Tyler/Co. D, W.S.S. (inscribed) [66th Ill. Inf.] | |
| 1692 | C. Parish, Co. F, 66th Ill., 2d Brig. 4th A.C. [listed] | |
| 1950 | Captured by Lt. Col. George A. Martin, 38th Va. Inf. (documented) | |
| 1978 | Pres. to Capt. J.H. Burton 1st Conn. H. Arty. by O.F. Winchester Oct. 23, 1863 (inscribed) | silver plated, frame engr. |
| 2045 | James E. Ramsey, Co. B., 40th Regt. Ind. Vols. (inscribed) | buckhorn sight |
| 2194 | Reported as captured from Indians 7-27-1879 | |
| 2274 | Pvt. Hiram S. Vinson, Co. K, 66th Ill. (reported from 1864 diary entry) | |
| 2287 | David Padgett Co. K/66th Ill. W.S.S. (inscribed) also marked L/341 | |

## The Historic Henry Rifle — Oliver Winchester's Famous Civil War Repeater

| No. | Identification | Remarks |
|---|---|---|
| 2293 | John Fox. 58th Ind. (reported - inscribed) | |
| 2317 | Pres. to Gov. W.F.M. Arny by E.M. Stanton Secy. of War, Aug. 1863 (inscribed) New Mexico Territory | |
| 2347 | John F. Phillips/58th Ind. (inscribed) | |
| 2352 | A.C. Holmes (scratched on stock) [Co. G, 10th West Virginia Inf., '62–'64] | |
| 2378 | Major General J.G. Blunt (inscribed) appt'd. Maj. Gen. 5-'63 | gold plated, engraved |
| 2457 | J.B. Martin 11th Ky. Cavl. Co. M 3rd Battn. (inscribed) | buckhorn sight |
| 2514 | Levi Brown, Co. E. 13th Mich. V.V. (double inscribed, but first name crossed out) James H. Smith Co. H, 13th Mich. V.V.I | no rear sight |
| 2523 | J.W. VanBrocklin/Co. D, 66th Ill. Western Sharpshooters - Res'd. Battle Creek, Mich. (inscribed, also from diary entry 1864) | |
| 2545 | Pvt. William P. Beesley, Co. I, 66th Ill. W.S.S. - (from diary entry 1864) | |
| 2575 | Sergt. L.A. Barker/Co. D, 66th Ill. Vet. Vols. W.S.S. - [purchased at Corinth, Miss. Sept. 4, 1863, for $40] (inscribed after war) | |
| 2582 | L.P. Tallman/Co. A. Western Sharp Shooters Vet. Vols. (inscribed) | |
| 2637 | Pvt. Daniel W. Litherland, Co. I, 66th Ill. W.S.S. (died 4-10-1864, from diary entry) | |
| 2639 | Chas Webster/Co. D, W.S.S. Vet. Vols. (reported inscribed) | |
| 2688 | Thomas E. Gleason. Co. D, W.S.S. (inscribed) [died of wounds 11-7-1864] | |
| 2710 | Captured at Goodrich's Landing, Oct. 1864 J.G.S. - G.M.S. Pauline (inscribed) | |
| 2724 | A.J. Rolf/Co. D/23d Ill. Inf. (inscribed) | |
| 2729 | reported carried by High Backed Wolf - Cheyenne warrior K.I.A. Platte Bridge, Oregon Trail 7-25-65 | |
| 2779 | John Cowan, Co. D, 23rd Ills. Vols. (inscribed) | |
| 2797 | associated with Co. F 97th Illinois Inf. per National Archives records | |
| 2857 | Henry V. Hoagland, Co. F, 7th Ill. Vet. Vol. Inf. (inscribed) | |
| 2900 | J.M. Bodge. Co. H, 66th Ill. W.S.S. Vet. Vols. (inscribed) | |

*Appendix D*

| No. | Identification | Remarks |
|---|---|---|
| 2909 | J.A. McClure Co. D, 57th Regt. Ind. V.V. (inscribed) [KIA 6-28-64] | |
| 3117 | associated with Co. K, 7th Illinois Inf. per National archives records | |
| 3183 | J.K. Watson/64th Ill., Co. G (inscribed) | |
| 3233 | associated with Co. F, 97th Illinois Inf. per National Archives records | |
| 3237 | associated with the 97th Ill. Inf. | |
| 3261 | owned by R.H. Bates, Co. A, 29th Texas Cav. CSA March 24, '65 (inscribed) | |
| 3315 | Lockwood Sanford (inscribed) | |
| 3347 | associated with Co. F, 97th Illinois Inf. per National Archives records | |
| 3406 | J.C. Miller, U.S. Arsenal Aug. 24, '66 L.C. Stockwell Davenport, Iowa, T.D. James, Mar. 18, 1868; John Magill, Sept. 1868 (inscribed) | US-CGC |
| 3469 | Pres. to Private David Reed by the U.S. Gov't. Mar. 2, 1865 (plaque) [3rd U.S.V.V.] | US-CGC |
| 3509 | associated with Co. F, 97th Illinois Inf. per National Archives records | |
| 3532 | associated with Co. F, 97th Illinois Inf. per National Archives records | |
| 3930 | Bangor, Maine/Pvt. S.A. Holway Co. H 1st Dist. Col. Cav. (inscribed) [bbl. no. restamped] | US-CGC |
| 4140 | J.D. Orcutt Co. A, 7th Ill. V.V.I. (inscribed) [reg't. rec'd. Henrys July 27, 1864] | |
| 4178 | G. Burkhardt Co. H, 7th Ill. Vol. Inf. (reported inscribed) | |
| 4228 | associated with Co. I, 73rd Illinois Inf. 11-08-1864, per National Archives records | |
| 4392 | C.M. Baker, Co. E, 36th Ill. (reported as inscribed) | |
| 4434 | M.G. Buzard/Co. D ...log's Eng's Mo. Vols. (consol. with 25th Mo. 2-4-64) (inscribed) | |
| 4494 | Nicholas Ham, Co. F, 64th Ill. Vet. Vol. (inscribed) also reported as #4944 | |
| 4615 | Lt. D.M. Freman/Missouri Vol. Militia (inscribed) | |
| 4658 | Sgt. E.A. Moore, Co. E, 10th Ill. V.V. Inf. (inscribed inside the sideplate) | |
| 4983 | Pres. to Lt. William J. Creasey by the citizens of Newburyport, Mass., 1867 (inscribed) | |

# The Historic Henry Rifle — Oliver Winchester's Famous Civil War Repeater

| No. | Identification | Remarks |
|---|---|---|
| 5095 | S.W. Vestal, Wilmington, Ohio [40th Ohio Inf.] (inscribed) | |
| 5218 | Col. Eli Long/4th Ohio Cavalry (inscription not authenticated) | engraved |
| 5594 | Marvin D. Bowin, 39th Ill. Vet. Vol. Inf. Chicago, Ill., Oct. 20, '64 by his uncle E.R. Bowen (inscribed) [K.I.A. 4-2-1865] | |
| 5952 | Capt. James M. Wilson, Co. B, 12th Kentucky Cav. (inscribed) | |
| 6039 | John L. Batchelder from Henry A. Morse (inscribed) | |
| 6083 | Henry Parkhurst (inscribed) | |
| 6248 | De Leuw (inscribed) [assoc. with the 7th Ill. Inf.] | |
| 64xx | Capt. Hyman Co. D, 115th Ill. Inf. (reported inscribed) | |
| 6639 | inscribed to soldier in 115th Ill. (reported) | |
| 6699 | Walter Passavant-Pittsburg, Penna. (inscribed) | |
| 6854 | Durbin to Shaler [said to be rifle of Brig. Gen. Alex. Shaler – 6th Army Corps] (inscribed) | |
| 7136 | Sitting Bull from the President of the U.S. (inscribed) | |
| 7169 | Capt. Edward J. Merrill/by the Board of the U.S. Draft Rendezvous March 1865 (inscribed) | |
| 7225 | G.W. Fulton (inscribed) | engraved |
| 7343 | F.D. Pease, Bloomfield, Ind. [Pvt., Co. B, 1st Mich. Inf., later U.S. Vet. Vols.?] (inscribed) | |
| 7365 | J. Spangenberg, Co. B, 3rd Regt. U.S.V.V. | (inscribed) |
| 7493 | John Anderson, Co. H, 3rd Regt. U.S.V.V. (recorded, with documentation) | |
| 7535 | D. Bolton (inscribed) | |
| 7582 | Private Jacob Werle, Co. B, 3rd Regt. U.S.V.V. (inscribed, with documentation) | |
| 7733 | J.M. Leader, Co. E, 3rd Regt, 1st A.C., U.S.V.V. Pres. from the city of Phila., Pa. (inscribed) | |

# Appendix D

| No. | Identification | Remarks |
|---|---|---|
| 8051 | Frank W. Meese/Kane, Pa. (inscribed) | |
| 8622 | T. Pfeifer, Co. B, 3rd Regt., 1st AC. U.S.V.V. | (right frame) |
| | Amos F. Moore, Palo, Ill. (left frame) (inscribed) | |
| 8862 | reported as captured from Indians 7-27-1879 | |
| 8909 | with Japanese characters, sold in 1868 to Emperor Meijo by Paul Barnett (reported) | |
| 8929 | Abraham Miller, Jr. (inscribed) | |
| 8972 | reported as captured from Indians 7-27-1879 | |
| 9193 | George W. Petit, 3rd Regt. [U.S.] 1st A.C. Veterans (inscribed) | |
| 9223 | Archd. McAluster, Co. E, 2nd Regt. P.R.V.C.; enlisted Apr. 27, 1861, discharged Jun. 16, 1864, re-enlisted Mar. 17, 1865 Co. H, 3rd Regt. Hancock's 1st A.C., discharged Mar. 16, 1866 (inscribed) [3rd U.S.V. Vols.] | |
| 9373 | Pres. to Capt. John A. Smith, Co. E 7th Regt., Ill. Inf. by the troops of his command for outstanding leadership and valor (inscribed) [mustered out 7-25-1865] | converted to .44 cf. |
| 10169 | Jacob Rideout, Contention, Arizona Territory (inscribed) | |
| 10547 | O.F. Winchester, New Haven, CT, U.S.A. (inscribed) | fitted with magazine cut-off device, 9-4-'66 patent |

## Inscribed Henry Rifles With No Serial Numbers Reported

| | | |
|---|---|---|
| xxxx | Sgt. E.A. Putnam, Co. D, 9th N.Y. Cav. | |
| xxxx | G.W. Coffman, Co. B, 6th W.V.V. Cav'l. W. Rollins, Co. B, 6th W. Va. Cav'l. (crude inscription) | |
| xxxx | "S.H. Thompson, Co. G, 80th Ills." (inscribed) | |
| xxxx | "Maj. J.C. Smith, 5th O.V.C." (inscribed) engraved with deluxe stock | |
| xxxx | "Col. Z.R. Bliss" (inscribed) [Col. 10th & 7th R.I. Inf.] non-factory engraved | |
| xxxx | Sgt. R.L. Schick, 1st Dist. of Columbia Cavalry, Maine, "Death to Traitors" (crudely inscribed) | |
| xxxx | (Presented to Sgt. Gilbert Armstrong of Co. E, 58th Indiana for bravery at Stones River) this rifle was captured at Chickamauga, where Armstrong was wounded | |
| xxxx | "Lieut. S. Ames/3d K.S.V." (reported to be inscribed) | |
| xxxx | "A.C. Holmes/10th Regt. W.V. Inf. Vols." (inscribed) | |

# APPENDIX E

## Comments Regarding the Iron-Framed Henry Rifle

Speculation has long existed regarding the New Haven Arms Company's early use of iron as well as brass frames and buttplates. Due to the early difficulties and lengthy delay in obtaining equipment and also tooling-up for Henry rifle production (largely due to the need to obtain special machinery with which to make the much larger component parts of the Henry versus the old Volcanic arms), Oliver Winchester may have contracted with an outside firm (such as Colt's) to supply an initial quantity of iron frames and buttplates to help expedite production of the Henry rifle. This is consistent with other parts procurement by the New Haven Arms Co., including 2,500 "finger levers from our patterns" (Arcade Malleable Iron Co., ordered 3-25-1863, letter #264). Otherwise, Oliver Winchester may have considered iron frames to be important in a potential U.S. Navy sale. Considering that some other arms manufacturers (including Colt's with its M1851 iron-strapped pistols) had replaced brass furniture with iron for U.S. Navy sales due to the tarnishing effect of a salt water environment, which required brass to be frequently polished, the prospect of Navy use may have been foremost in Winchester's mind when he initiated the manufacture of about 200-plus iron-buttplated and iron-framed rifles (see text, part II). Yet, the iron-framed Henry did not prove of sufficient merit to warrant its continuance, especially as no direct U.S. Navy sales were forthcoming. These iron-framed rifles, which may have been intended as a separate entity due to their special purpose, were soon sold commercially along with the brass-framed rifle. That the brass frame became standard was economically justified by the reduced raw material cost and ease of machining with non-ferrous metal. Also, the brass frame was of slightly lighter weight, while being sufficiently strong. In all, the brass was more practical to manufacture and quite user friendly.

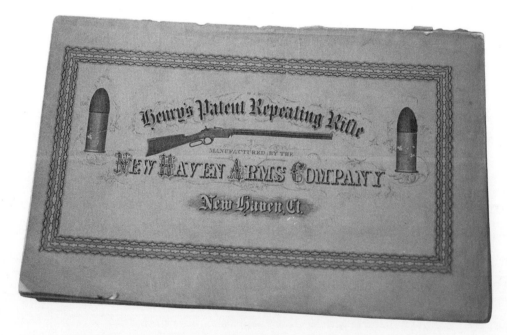

**New Haven Arms Company 1865 catalog.** (Photo Courtesy Les Quick)

## Iron and Brass Frame Serial Numbers

From surviving rifles it is believed about 200 to 300 iron-frame Henry rifles were manufactured at the start of production (April/May/June 1862). Although there have been some reported duplications of iron-frame and brass-frame serial numbers, it is my estimation that the iron-frame and brass-frame numbers are intermixed within the same serial number sequence. A duplication of serial numbers would represent a nightmare of accountability for the factory, especially since serial numbers were utilized to keep track of defects and repairs. Also, the New Haven Arms Co. used their serial number ledger to determine where and to whom each rifle was shipped. This proved to be of importance in attempting to learn where B. Kittredge & Co. was obtaining their Henrys in 1863. Furthermore, considering the factory's known intermixing of Henry and Improved Henry (Model 1866) serial numbers in 1867, there was no precedent or other practical basis to explain why the factory would establish separate iron and brass-frame serial numbers. An occasional duplication in numbers can be better explained by some rifles being stamped with the same number due to early in-plant contractor confusion (especially if the iron-frame arms were originally processed in separate factory lots expressly as a Navy rifle) and/or mistakes prior to the normal pattern of production. That the factory was less than perfect with their rifle serial number accountability is witnessed by Letter Book entry #103, dated October 10, 1862, wherein agent John W. Brown was told to send a list of rifle numbers from his last shipment, as "we omitted to take them up on our books." Such factory carelessness may have resulted in various duplicated numbers, especially due to the lot or batch production.

## List of Reported Iron-Frame Henry Rifles

Please note: the following list of iron-framed Henry rifles is neither complete nor likely to be totally accurate. These are rifles that have been reported as having iron frames and iron buttplates, an early manufacturing alternative that was soon discontinued. There is no assurance that the serial numbers reported are accurate. This list is presented merely as a compilation determined from a variety of sources, including dealer lists, published arms, personal observation and collector lists. A special thanks is due to Jim Gordon, Sante Fe, NM, and Doug Bennick, Orange, CA, for the use of their historical data and lists. Howard Madaus, who considers the iron-frame rifles to be of a separate production and serial number sequence, provided additional valuable information.

| **2** (41) | 49 | **66** | 89  | 108 | 129 | **149** | 167 | 197     | 270 | 289     |
|------------|----|--------|-----|-----|-----|---------|-----|---------|-----|---------|
| **12**     | 51 | 68     | 90  | 110 | 131 | 152     | 170 | **205** | 275 | 304     |
| 13         | 54 | 70     | 94  | 114 | 134 | 155     | 175 | 209     | 276 | 323     |
| **20**     | 55 | 71     | 95  | **117** | 135 | **156** | 177 | **215** | 277 | **324** |
| **31**     | 57 | 73     | 103 | 119 | 137 | 157     | 179 | 257     | 278 | **355** |
| **43**     | 59 | **78** | 104 | 123 | 138 | 159     | 182 | 260     | 279 |         |
| 45         | 62 | 85     | 105 | 125 | 147 | 161     | 185 | 262     | 281 |         |
| 48         | 64 | 88     | 106 | 128 | 148 | 165     | 192 | 267     | **287** |     |

**TOTAL REPORTED 85**

Etimated number of iron-framed rifles manufactured — 203.

Duplicate iron and brass-framed serial numbers reported (15):

**2, 12, 20, 31, 43, 66, 78, 117, 149, 156, 205, 215, 287, 324, 355**

# The Historic Henry Rifle — Oliver Winchester's Famous Civil War Repeater

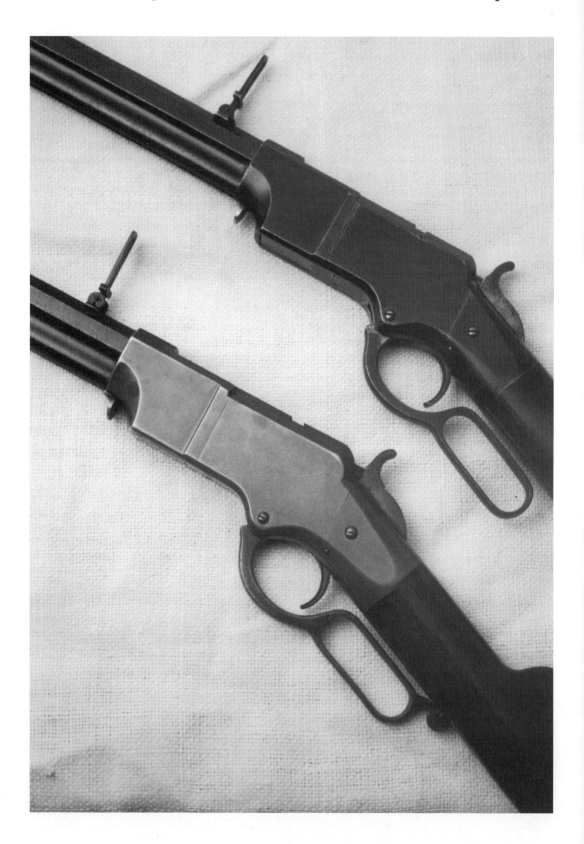

*Appendix E*

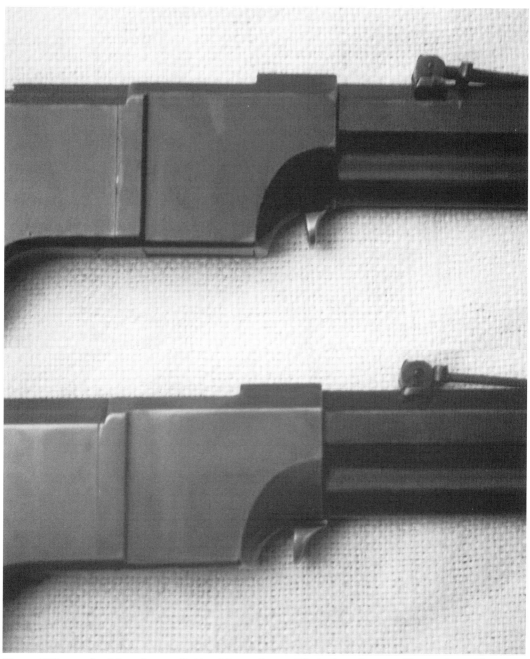

**Henry Rifles —** *(top)* **iron frame,** *(bottom)* **brass frame. Photo is a close-up illustration of iron vs. brass.**  (COURTESY OF LES QUICK)


FACING PAGE

*(top)* Iron frame.
*(bottom)* Brass frame.
**Photo illustrates differences in frame material used during production.**  (COURTESY OF LES QUICK)

# APPENDIX F

## Basic Production Configurations of the Henry Rifle

There were four major production changes (variations) in the Henry rifle during the course of its production from the spring of 1862 to the last shipment ca.1867. Generally, they involve the frame configuration or construction.

1. Iron frame, slotted for a rear sight, and round heel *(top)* pattern iron buttplate, no lever locking latch. Serial nos. intermixed with brass-framed production, ser. nos. 2 to 355 reported, estimated total production about 200.
2. Brass frame slotted for a rear sight, but most rifles carried only barrel-mounted sights. Round heel-style brass buttplate. Lever locking latch standard from about ser. no. 325. Serial no. range 1 to apx. 3000.
3. Plain brass frame, (not slotted for a rear sight). A minor variation of this configuration occurred when the factory adopted the pointed heel-style buttplate during the 4000 serial number range. Plain (unslotted) frames began about serial no. 3000 (coinciding with the U.S. martial serial number primary range).
4. Smooth rounded brass frame; the sharp corners of the top of the frame were rounded near the cartridge ejection port. These were among the last few Henrys manufactured; all post Civil War.

Other minor changes occurred in stock (comb) profile: diameter of cartridge slot opening in the frame; pattern of firing pin and bolt; diameter of cleaning rod port in the buttplate, and the use of rivets versus screws in sling swivels mounted on the barrel and stock (an optional feature).

Photo illustrating the two basic differences in buttplate configuration:

*(top)* early configuration with rounded upper portion; *(bottom)* later style with longer pointed upper extension.

(COURTESY LES QUICK)

*(Right)* **Barrel sling swivel fitting.**

*(Below)* **Comparison of serial number marking. Rear sight locations and brass patina. Top rifle, sn7644, middle rifle, sn1677, bottom rifle, sn14.**
(BOTH PHOTOS COURTESY LES QUICK)

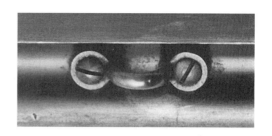

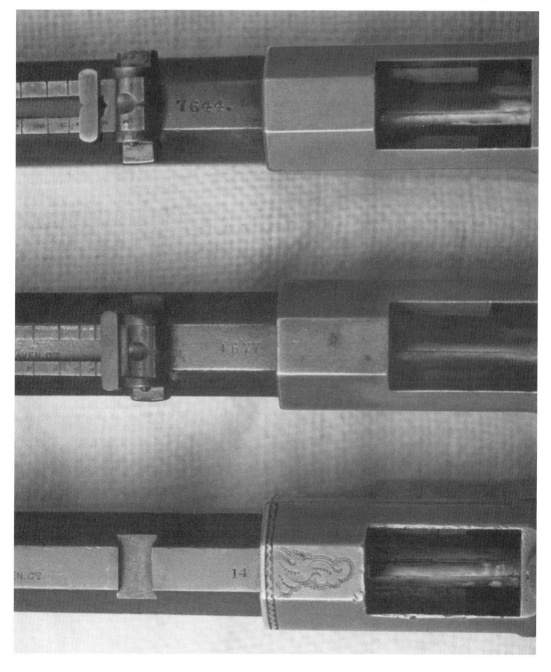

# U.S. GOVERNMENT TRIALS 1864

**TRANSCRIPTION OF HENRY RIFLE DOCUMENTS
FROM NATIONAL ARCHIVES RECORDS: RG 156 E 201 p. 414**

**OLIVER WINCHESTER COMPLAINS ABOUT THE INITIAL HENRY CARBINE TEST**

New Haven [Ct.] Apl. 20th 1864

A.B. Dyer
Major [of] Ordnance, Comdg. Springfield Armory

Sir:

Your favor of the 2d instant, covering a copy of your report of the trial of the Henry Repeating Carbine [dated March 30, 1864], submitted by me as the representative of the New Haven Arms Co., was duly received.

As this report is strongly adverse to the arm, and as the incidents are not correctly stated from which the conclusions were drawn, and thus an injustice is done to us & the weapon (as I am not willing to believe that you purposely do us such injustice) I will take the liberty to review the report.

"The arm became unserviceable three times; once in consequence of the disarrangement of some of the parts, once by the breaking of a spring, and once by the bursting of a cartridge case; the latter is liable to occur frequently."

You are correct in the two first facts, but the remedy is simple & sure to prevent a possible reoccurrence of them.

The third fact was not the fault of the gun, nor is it any more liable to occur with this, than any other gun. Nor did it render the gun unserviceable, nor did it involve the necessity of taking the gun apart as stated in Lieut. Smoot's and Supt. Allin's report. Had I been present, I could have shown you how to have removed both cartridges in a few seconds without removing a single part of the gun. This model was furnished only a day before it was presented, and without other trial than firing six charges. We expected that if there were any parts too weak for the duty required of them, they would be developed [discovered] in the trial. To insure this, we procured cartridges from Leete & Co. that had been made for you of a length to hold forty grains of powder. These we reduced in length so as to hold but thirty seven grains. As you use them in these [tests], we compressed [into them] 45 to 55 grs. of powder. [But] the heads of the cartridges proved too thin to stand the greatly increased force of these charges (we use thicker copper in ours — [no.] 42 carrying 25 grains). Consequently the head of the cartridge was separated entirely from the cartridge (as shown by Mr. Porter) under such circumstances thereby [whereby] no cartridge drawer [extractor] in any gun that [is made] would have withdrawn the shell. Yet for this, you would make the gun responsible, & add that it is likely to occur frequently.

I have known millions of cartridges to be fired with the same arrangement in our other gun [standard Henry rifle] & never knew this to occur before, or it to once fail of withdrawing the charge or shell. Again it is stated that in consequence of this bursting of the shell, it was necessary to take the gun apart & that it is a difficult gun to take apart, requiring half an hour.

In reply I would say that the time taken (like the supposed necessity of taking it apart at all) is not the fault of the gun, but [a] want of familiarity of your men with it. I set three of our men to strip these guns [apparently the standard Henry rifle] entirely apart, & reassemble them. The average time required was less than nine minutes. "It cannot be loaded except at full cock, & it has no half cock or safety notch." True, but it can have if wanted. We claim, however, & hundreds who have used it much admit that it is much better as it is. "Several times the gun worked very hard." Not so. The appearance of this was due to its being held in a frame that was not suited to the form of the gun, so that the lever could not be taken hold of in the proper place. [Earlier, in my factory] I did all the loading & had to do it with one finger & then work the lever & that close to the barrel & fulcrum of the lever. "Once a ball came out of the cartridge." True, but this was one of the cartridges I first brought to Springfield & was <u>not</u> <u>fastened</u> into the shell, as I stated to you at the time. "Its mechanism is too delicate & complicated, & the cost to make [the carbine] is necessarily very great." The mechanism can be made strong enough for all the requirements of the service without increasing the weight, & we can furnish them of the same price the government is paying for other [repeating] carbines. "The magazine seems liable to be injured." It is of solid steel, & stiffer than a Springfield musket barrel. "It (the tube, or magazine) is likely to get lost." So is the magazine of the Spencer Carbine, the ramrod of any gun, and <u>other</u> <u>appendages</u> which this gun does not require. But if it does get lost & all the repeating arrangements <u>ruined</u>, it can still be used as a single shooter with as much facility as any other breechloader. "It weighs 8 lbs. 3 ozs., which is a pound & a half more than it should be." It weighs without appendages (cleaning rod) but 7 lbs. 14½ ozs. A similar gun weighing 9 lbs. 14 ozs. has been in use in the Cavalry service nearly two years [i.e. standard Henry Rifle] & the weight has never been objected to, nor would the men or officers exchange them for any other. Thus, as it appears to me, there is not a single sound exception or objection taken to this gun in the report, which cannot be easily & completely removed. My mistake was in not thoroughly testing & correcting such weak points as were developed, before submitting it for trial. I mistook the spirit & manner in which I supposed such trials to be conducted. I supposed it was the interest of the Government to foster & encourage every effort to improve & perfect weapons of self defense, & if a gun had one desirable feature in it, such as this has — that of <u>loading</u> <u>with</u> <u>great</u> <u>rapidity</u> — that feature would receive some notice, & with some encouragement to remedy the minor defects that might appear. I observe also that your report is marked "final." Please explain if that means no further trial can or need be had. Trusting that you will see the force of the exceptions taken to your report, I will so far as is right & proper, take steps to correct any wrong it may otherwise do.

I am, very respectfully yours,<br>
O.F. Winchester<br>
Prest., N.H. Arms Co.

# The Historic Henry Rifle — Oliver Winchester's Famous Civil War Repeater

**MAJOR DYER STANDS HIS GROUND**

Springfield Armory
April 26, 1864

Brig. Genl. Geo. D. Ramsay:
Chief of Ordnance,
Washington, D.C.
Sir:

    I have the honor to transmit herewith a copy of a letter I have received from Mr. O.F. Winchester, Prest. Of the New Haven Arms Company, reviewing my report on the Henry Repeating Carbine, dated March 30th, 1864, a copy of which was furnished him by your direction.

    I have no reason to change in any respect my report.

Very respectfully,
Your obt. Svt.
A.B. Dyer,
Maj. of Ord., Cmdg.

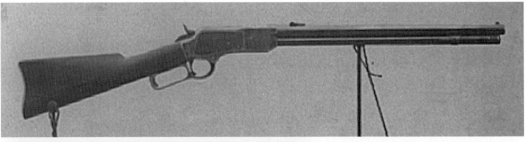

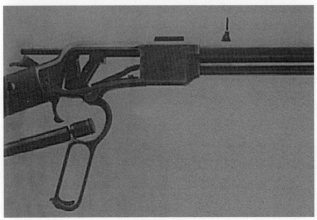

Official U.S. Army photographs of the Special Test Henry Carbine submitted for evaluation in August 1864. Note the resemblance in profile to the much later M1873 Winchester. This is believed to be the exact same carbine submitted in January. Both trials failed, and Oliver Winchester was thwarted in his attempts to sell carbines to the army. At left is exploded view of this experimental carbine. Note the unique forward-loading tubular magazine design.
(COURTESY OF THE NATIONAL ARCHIVES, WASHINGTON, D.C.)

## U.S. Government Trials 1864

**TRANSCRIPTION OF HENRY RIFLE DOCUMENTS
FROM NATIONAL ARCHIVES RECORDS: RG 156 E 201 p. 468
REPORTS OF EXPERIMENTS 1826-1871**

**REPORT ON HENRY CARBINE, MADE BY LIEUT. H. STOCKTON
OF THE WASHINGTON, D.C., ARSENAL, OCTOBER 1864**

October 26th 1864

Major J.G. Benton
   Comdg. Arsenal
Major:

In accordance with your instructions, I have examined and tested Henry's Carbine and have the honor to submit the following report.

Description:

The carbine is a modification of Henry's Rifle, which it closely resembles. The barrel is shorter, however, and is rifled with a uniform twist of one turn in 27". Under it is a magazine of nearly the same length, composed of a tube of thin steel, capable of containing nine cartridges. The bottom cartridge, which rests on a carrier block in a sort of chamber behind the magazine is brought on a level with the bore by a forward motion of the trigger guard, which at the same time brings the hammer to a full cock. On forcing the guard back again to its place, the cartridge is thrown forward into the chamber, the carrier block returns to its place and receives another cartridge from the magazine, and the piece is ready to fire.

A spiral spring is inserted into the magazine to press the ammunition into the receiver. This is withdrawn on loading, just as a ramrod would be, by springing the head out from a notch near the muzzle. The caliber is: .44. Diameter of chamber: .455. Depth of grooves: .075. Weight of ball: 324 grains. Weight of powder 36 grains. The cartridges are copper case percussion and are exploded by the action of the hammer on a horizontal steel cylinder, with two points, which are made to strike the edge of the case. After firing, the case is withdrawn by this same cylinder, and expelled with considerable force from the piece.

   The advantages claimed for this carbine are:
   1st  Rapidity of fire.
   2d  Superiority over other magazine carbines in that even if the spring acting on the cartridges be lost, the working of the piece will not be affected if the muzzle be kept slightly elevated.
   In other respects it is claimed to be fully equal to any other carbine.

## TRIAL

The piece was tried in the presence of Mr. Winchester, president of the New Haven Arms Co.

1st  Accuracy of fire (see target record). Distance to target 300 yards. Wind fresh from front & rapid. Three shots were fired to get range. 20 shots with elevation for 350 yards. Three of the cartridges had to be struck twice with the hammer before they exploded.

2nd 20 shots at 500 yards, elevation for 550, piece wiped out before firing. Five cases were too thick and had to be struck twice.

3rd Commenced to fire for rapidity and endurance, but the piece becoming foul, began to work very badly, and at last became entirely useless.

At this stage of the proceedings Mr. Winchester requested that the further trial be deferred until he should have made some alterations to prevent the fault just mentioned. This was on the 26th of August and since then nothing has been heard from Mr. Winchester. The report is therefore submitted in the present incomplete form.

During the trial it not infrequently happened that the cylinder hooks failed to extract the old cartridge case, in which event the new cartridge was jammed up behind it, not being able of course to enter the chamber. The remedy for this lies in pressing down the carrier block and new cartridge with the finger, when as soon as the latter is brought to the original position, the old case may be extracted. Of course, this occasions considerable detention.

It is still worse when a cartridge fails to explode, and in attempting to withdraw it, the ball becomes detached from the case. The new cartridge must then be pressed down and a ramrod used to expel the ball.

A serious objection to the carbine is the fact that it has no half cock. With a load in therefore, it must be carried carefully. The arrangement of the lock would admit of this fault being corrected without much trouble.

> Very respectfully, your obt. Servt.
> H. Stockton
> 2nd Lt. Ord.

Respectfully forwarded to the Chief of Ordnance pursuant to his instructions of Sept. 10th 1864.

> J.G. Benton
> Maj. Ord. Cmdg.

# ENDNOTES

## PART I

1. New Haven Arms Company, Letter Book, May 1, 1857 to March 28, 1859, and October 8, 1862 to December 12, 1863, McCracken Research Library, Cody Firearms Museum, Buffalo Bill Historical Center, Cody, WY, (hereafter cited as NHAC Ltr. Bk.) entry #118, O.F. Winchester to E.B. Martin, October 18, 1862.
2. Parsons, John E., *The First Winchester, The Story of the 1866 Repeating Rifle*, New York, 1955, 1969, pp. 8–9.
3. NHAC Ltr. Bk. #118, 161, 334, 457, 473.
4. NHAC Ltr. Bk. #118.
5. *Louisville Journal*, June 26, 1862, July 14, 1862; Parsons, *First Winchester...*, pp. 13–14; NHAC Ltr. Bk. #138.
6. *Louisville Journal*, July 10, 1862, Aug. 1–30, 1862.
7. NHAC Ltr. Bk. #126, 138, 142; *Louisville Journal*, Aug. 30, 1862.
8. NHAC Ltr. Bk. #118, 121.
9. Private William Farries, 24th Wisconsin Inf., Sept. 27, 1862, author's collection; NHAC Ltr. #126, 132, 138, 253.
10. War of the Rebellion: Official Records of the Union and Confederate Armies, Washington, D.C., series 1, vol. 16, part 1, pp. 864, 867 [hereafter cited as OR].
11. NHAC Ltr. Bk. #108, 110.
12. NHAC Ltr. Bk. #97, 98, 99, 100, 106, 108, 117.
13. NHAC Ltr. Bk. #106.
14. NHAC Ltr. Bk. #100, 106, 109, 110, 130.
15. NHAC Ltr. Bk. #110, 121, 130, 136, 383.
16. NHAC Ltr. Bk. #47, 102, 124, 211, 266, 324.
17. Parsons, *First Winchester*, 1969 edition, addenda; Henry's Patent Repeating Rifle, New Haven Arms Co. Catalog, New Haven, Ct., 1865, p. 37 ff.; NHAC Ltr. Bk.#165.
18. NHAC Ltr. Bk. #166.
19. Parsons, *First Winchester...*, p. 23.
20. NHAC Ltr. Bk. #148, 150, 151, 153.
21. NHAC Ltr. Bk. #183.
22. NHAC Ltr. Bk. #200, 285, 338, 339.
23. NHAC Ltr. Bk. #165, 166, 168, 169, 172.
24. NHAC Ltr. Bk. #166, 171, 172.
25. *Louisville Journal*, Oct. 28, 1862; Henry 1865 Catalog, pp. 34–35.
26. NHAC Ltr. Bk. #117, 133.
27. NHAC Ltr. Bk. #133, 137, 140, 142.
28. NHAC Ltr. Bk. #160.
29. NHAC Ltr. Bk. #164, 198.
30. NHAC Ltr. Bk. #127, 182, 194, 198.
31. NHAC Ltr. Bk. #176.
32. NHAC Ltr. Bk. #178, 179, 198, 235, 315.
33. NHAC Ltr. Bk. #162; Parsons, *First Winchester...*, p. 9.
34. NHAC Ltr. Bk. #136, 156.
35. NHAC Ltr. Bk. #179.
36. NHAC Ltr. Bk. #162, 182, 261.
37. NHAC Ltr. Bk. #233.
38. NHAC Ltr. Bk. #259, 288, 292, 303, 319, 405, 419, 468, 480.
39. NHAC Ltr. Bk. #221.
40. NHAC Ltr. Bk. #173, 174, 187, 198.
41. NHAC Ltr. Bk. #184, 186, 187, 188, 192, 245.
42. NHAC Ltr. Bk. #245, 251, 293.
43. NHAC Ltr. Bk. #251, 322, 323, 330, 346, 433.
44. NHAC Ltr. Bk. #266; Parsons, *First Winchester...*, p. 12.
45. NHAC Ltr. Bk. #268.
46. NHAC Ltr. Bk. #203, 371.
47. NHAC Ltr. Bk. #252, 293, 306, 307.
48. NHAC Ltr. Bk. #268.
49. NHAC Ltr. Bk. #251, 253.
50. NHAC Ltr. Bk. #243, 255, 258, 259, 301, 346.
51. NHAC Ltr. Bk. #223, 268, 269, 320, 332.
52. NHAC Ltr. Bk. #197, 253, 328.
53. NHAC Ltr. Bk. #197, 208, 279.
54. NHAC Ltr. Bk. #102, 112, 130, 135, 228, 290, 381, 395, 466.
55. NHAC Ltr. Bk. #228.
56. NHAC Ltr. Bk. #171, 198, 276, 278, 286, 315, 320.
57. NHAC Ltr. Bk. #346.
58. Henry 1865 Catalog, p. 16; NHAC Ltr. Bk. #217, 220, 238.
59. NHAC Ltr. Bk. #312, 314, 323, 346, 350, 353.
60. NHAC Ltr. Bk. #427, 433.
61. NHAC Ltr. Bk. #376, 458.
62. NHAC Ltr. Bk. #472.
63. NHAC Ltr. Bk. #173, 477.
64. This data is based upon ammunition issues and historically identified Henry rifles. OR ser. 1, vol. 38, pt. 1, p. 126.
65. Henry 1865 Catalog, pp. 26–27, 35–36.
66. Henry 1865 Catalog, p. 32.
67. Henry 1865 Catalog, pp. 24–26.
68. Henry 1865 Catalog, pp. 20–23.
69. Henry 1865 Catalog, pp. 19, 27.
70. Henry 1865 Catalog, p. 5.
71. Henry 1865 Catalog, pp. 6–7.
72. Henry 1865 Catalog, pp. 7–8.
73. NHAC Ltr. Bk. #146, 225, 249, 266, 385,

395, 437, 482; Records Group 156, Entry 21, boxes 291, 292, letters 137L, 324W, 236WD, National Archives, Washington, D.C.
[74] Henry 1865 Catalog, pp. 9–10.
[75] *Ibid.*; general entries NHAC Ltr. Bk.
[76] Parsons, *First Winchester...*, p. 44; James D. Julia Auction Catalog, Fairfield, ME. April 26–28, 1999, item no. 98 (Henry rifle no. 10294, 22" barrel); RG 156 E 201, Reports of Experiments 1826–1871, Ordnance Dept., Vol. 9, #414, #468.
[77] Parsons, *First Winchester*, pp. 50–57.
[78] Parsons, *First Winchester...*, pp. 46, 49; Harold F. Williamson, *Winchester The Gun That Won the West*, New York, 1952, p. 42.
[79] Williamson, *Winchester...*, pp. 52–54.
[80] Williamson, *Winchester...*, pp. 40, 49.
[81] Williamson, *Winchester...*, pp. 42, 464.
[82] Parsons, *First Winchester...*, pp. 54, 58.
[83] Parsons, *First Winchester...*, p. 61.

# PART II

[1] Records Group 156–994 Correspondence Relating to Inventions, National Archives, Washington, D.C. (Hereafter cited as NA).
[2] Records Group 156 E3, Correspondence of the Chief of Ordnance, 7W2 vol. 53, p. 149, NA.
[3] *Henry Repeating Rifle Catalog*, New Haven Arms, Co., Hartford, CT., 1865, printed in L.D. Satterlee, *Ten Old Gun Catalogs for the Collector*, Chicago, IL., 1962, pp. 30–31.
[4] War of the Rebellion: Official Records of the Union and Confederate Armies, Washington, D.C., series 1, vol. 3, part 1, page 733 ff. [hereafter cited as OR].
[5] Oliver F. Winchester to E.B. Martin, Oct. 19, 1862, entry #118, New Haven Arms Company, Letter Book, May 1, 1857, to March 28, 1859, and October 8, 1862, to December 12, 1863, McCracken Research Library, Cody Firearms Museum, Buffalo Bill Historical Center, Cody, WY, (hereafter cited as NHAC Ltr. Bk.), see also Harold F. Williamson, *Winchester, The Gun That Won the West*, New York, 1952, pp. 393–394, n. 10; RG 74–165 Records of Ordnance Contracts, Sharps Rifle Mfg. Co., NA.
[6] Henry 1865 Catalog, p. 30.
[7] RG 74-165, Records of Ordnance Contracts — Spencer Repeating Rifle Co.; see also John D. McAulay, *Civil War Breechloading Rifles*, Lincoln, RI, 1987, p. 47.
[8] For photos of these arms, see R.L. Wilson, *Winchester, An American Legend*, New York, 1991, pp. 11, 15; OR series 3, pt. 2, p. 412.
[9] NHAC Ltr. Bk. #182, O.F. Winchester to Brig. Gen. Alfred W. Ellet, Jan. 31, 1863.
[10] For information on Baker, see Jacob Mogelever, *Death to Traitors*, the story of Gen. Lafayette C. Baker, Lincoln's Forgotten Secret Service Chief, Doubleday Co., Garden City, NY, 1960.
[11] *Louisville Journal*, June 26, 1862; Henry 1865 catalog, op. cit., pp. 36–38.
[12] RG 156 E79 p. 70 NA; RG 156 E21, F747 W(1863) NA; NH Ltr. Bk. #301 6-24-1863.
[13] NHAC Ltr. Bk. #301 6-24-1863; #303 6-24-63; #318 7-6-63; #329 7-21-63.
[14] NHAC Ltr. Bk. #301 6-24-1863 Oliver Winchester to James Ripley, June 24, 1863, also in RG 156 E 21 F747W (1863); George D. Ramsay to James Ripley, July 22, 1863, RG 156 E21 F869W (1863); RG 156 E 3-7W58 p. 474, NA. See also Henry rifle serial numbers listed in *U.S. Martial Arms Collector and Springfield Research Service Newsletter*, No. 78, Oct. 1996, 78-11–78-12.
[15] Merrill, Samuel H., *The Campaigns of the 1st Maine and 1st D.C. Cavalry*, Portland, ME, 1866.
[16] NHAC Ltr. Bk. #290-291, 6-12-63.
[17] NHAC Ltr. Bk. #320, 7-11-63; #378, 9-10-63.
[18] NHAC Ltr. Bk. #392, 9-19-63.
[19] NHAC Ltr. Bk. #433, 10-26-63; Merrill, *History of 1st Maine Cavalry*, p. 320; *U.S. Martial Arms Collector and Springfield Research Service Newsletter*, No. 78, pp. 11; RG 156 E 3 v.60 p. 185, 2-1-1864, NA. Ordnance Dept. Purchases, U.S. House Exec. Doc. 99, Serial 1338, pp. 276–277, 843; Receipt of Oct. 11, 1863, 60 Henry rifles, by Col. L.C. Baker, Washington, D.C., RG 156 E 21, NA.
[20] NHAC Ltr. Bk. #463, 11-16-63; #468, 11-17-63.
[21] NHAC Ltr. Bk. #478, 11-21-63; #480, 11-27-63; #483, 11-30-63; #491, 12-5-63.
[22] NHAC Ltr. Bk. #491,12-5-63.
[23] Oliver Winchester to W.A. Thornton, 1-30-1864, Letters Rec'd. by the Chief of Ordnance, 1864, RG 156 E 21, NA.
[24] RG 156 E 79 p. 70, 12-30-1864; RG 156 E 13, Letters, Telegrams, & Endorsements Sent, Ordnance Dept., v. 2, 378; RG 156 E 21, O.F. Winchester to Geo. D. Ramsay, 1-4-1864, NA.
[25] Ordnance Dept. Purchases, U.S. House Exec. Doc. 99, Serial 1338, *op. cit.*, pp. 276–277, 843; Receipt of 6-19-1864, 50 Henry rifles, from Columbus, Ohio,

Washington, D.C., RG156 E21, box 292, National Archives.

26. Letter of C. Alexander Thompson, New Haven, Ct., Jan. 27, 1863[4] to L. Dodge, author's collection.

27. RG 156 E 21, Letters Received by the Chief of Ordnance, O.F. Winchester to Col. W.A. Thornton, 1-28-1864, NA.

28. RG 156 E 3, Corresp. of the Chief of Ordnance, Col. W.A. Thornton to New Haven Arms Co., 1-29-1864, NA.

29. RG 156 E 21, Letters Received by the Chief of Ordnance, O.F. Winchester to Col. W.A. Thornton, 1-30-1864; O.F. Winchester to Brig. Gen. Ramsay, 1-30-1864, NA.

30. RG 156 E 21, Letters Rec'd. by the Chief of Ordnance, F. Winchester to Brig. Gen. Geo. D. Ramsay, 2-2-1864, NA.

31. RG 156 E 21, Letters Rec'd. by the Chief of Ordnance, F. Winchester to Brig. Gen. Geo. D. Ramsay, 2-5-1864; RG 156 E 3, Corresp. of the Ch. of Ordn., Geo. D. Ramsay to New Haven Arms Co. 2-5-1864; RG 156 E 3 v. 60, pp. 192, 227, Geo. D. Ramsay to O.F. Winchester, 2-3-1864, 2-9-1864, NA.

32. RG 156 E 13, Letters, Telegrams, & Endorsements Sent, Ordnance Dept., v. 2 p. 378; Ordnance Summary Statements, RG 156 E 110 for the 1st D.C. Cavalry, 1st, 2nd, 3rd Qtrs, 1864, NA.

33. George D. Ramsay to Col. J.C. Killon, 8-3-1864, RG 94 1st D.C. Cavalry, National Archives. See also, John McAulay, *Civil War Breechloading Rifles*, Lincoln, RI, 1987, pp. 45–47.

34. Dyer, Frederick H., *A Compendium of the War of the Rebellion*, New York, 1959, vol. 3, pp. 1018–1019; see also Ordnance Summary Statements, RG 156 E 110 for the 1st D.C. Cavalry, 1st, 2nd, 3rd Qtrs, 1864, NA.

35. Senate Report No. 183, 42d Congress, 2nd Session, Serial No. 1497, Sales of Arms by the Ordnance Department, 167; U.S. House of Representatives, Document 89, 42d Congress, 2nd Session, Serial No. 1511, p. 9, Sales of Arms and Ordnance Stores, Springfield Research Service, Serial Numbers of U.S. Martial Arms, vol. 4, p. 300.

36. RG 156 E 201, Reports of Experiments 1826–1871, Ordnance Dept. vol. 9, #414, O.F. Winchester to Major A.B. Dyer, 4-20-1864, NA.

37. *Ibid.*

38. *Ibid.*

39. Brig. Gen. Geo. D. Ramsay to Peter H. Watson, 4-5-1864, RG 156 E 21, Corresp. of the Chief of Ordnance, NA; Williamson, *Winchester*, p. 347.

40. NHAC Ltr. Bk, #472 11-18-63; #491 12-5-1863.

41. RG 156 E 201, Reports of Experiments 1826–1871, Ordnance, Dept. vol. 9, #468, NA.

42. *Ibid.*

43. RG 156 E 1012, Reports and Correspondence of Ordnance Boards, Laidley Board of 1865, including final report dated April 6, 1865, Major T.T.S. Laidley, Springfield Armory, NA; Roy Marcot, *Spencer Repeating Firearms*, Irvine, CA, 1983, pp. 100–104.

44. RG 156 E 79, Contracts for Purchases of Small Arms, vol. 1, p. 70, 4-7-1865, 5-16-1865; RG 156 E 3 7W2 vol. 63, p. 428.

45. Records Group 156, E 111, Ordnance Summary Statements, 3rd U.S. Veteran Volunteers, NA; *U.S. Martial Arms Collector and Springfield Research Service Newsletter*, No. 78, Oct. 1996, pp. 78-11, 78-12.

46. *Ibid.*, John D. McAulay, *Civil War Breechloading Rifles, op. cit.*, p. 47.

# PART III

1. Fagen, Dan P., "The Dimick Rifles of the 66th Illinois Infantry," *The Gun Report*, March 1996, vol. 41, no. 10, 16 ff; Dan P. Fagen, "Swiss Chasseurs, Dimicks & the 66th Illinois Infantry," North South Trader's *Civil War*, vol. 23, no. 2, March/April 1996, pp. 26 ff.

2. *Ibid.*
3. *Ibid.*
4. *Ibid.*
5. *Ibid.*

6. Fagen, "The Dimick Rifles...," *Gun Report*, March 1996, p. 17; Richard A. Baumgartner and Larry M. Strayer, *A Brief History of the 66th Illinois Infantry, 1861–1865*, in Lorenzo A. Barker, *With the Western Sharpshooters, Michigan Boys of Co. D, 66th Illinois*, Burlington, WV, 1994, p. 162.

7. Strayer and Baumgartner, *A Brief History of the 66th...*, p. 162; Fagen, "The Dimick Rifles...," *Gun Report*, March 1996, p. 17.

8. New Haven Arms Co. Letter Book, May 1, 1857 to March 28, 1859, and October 8, 1862 to December 12, 1863, McCracken Research Library, Cody Firearms Museum, Buffalo Bill Historical Center, Cody, WY, (hereafter cited as NHAC Ltr. Bk.) entry #307, 437, 482; Parsons, *First Winchester*, p. 14.

9. Fagen, "The Dimick Rifles...," *Gun Report*, March 1996, p. 24.

[10] Strayer and Baumgartner, *A Brief History of the 66th...*, p. 164; John E. Parsons, *The First Winchester The Story of the 1866 Repeating Rifle*, New York, 1955, p. 14.
[11] Baumann, Ken, *Arming the Suckers*, Dayton, Ohio, 1989.
[12] Barker, *With the Western Sharpshooters*, p. 34.
[13] Strayer and Baumgartner, *A Brief History of the 66th...*, pp. 166, 167.
[14] Records Group 156, E 111, Quarterly Ordnance Summary Statements, 66th Illinois Infantry, 1863–1864, National Archives, Washington, D.C.
[15] Strayer and Baumgartner, *A Brief History of the 66th...*, 170; Fagan, "The Dimick Rifles...," *Gun Report*, March 1996, p. 24.
[16] Strayer and Baumgartner, *A Brief History of the 66th...*, p. 170.
[17] Document in the possession of Dan Fagen, Florissant, MO.
[18] Fagen, "The Dimick Rifles...," *Gun Report*, March 1996, p. 20.
[19] Prosper Bowe letter of July 28, 1864, is printed in Strayer and Baugartner, *A Brief History of the 66th...*, pp. 174–175.
[20] Barker, *With the Western Sharpshooters*, p. 19.
[21] Strayer, Larry M. and Richard A. Baurgartner, *Echoes of Battle The Atlanta Campaign*, Huntington, West Virginia, 1991, p. 174.
[22] William P. Chambers' Journal Diary, entry for October 4, 1864, 46th Mississippi Infantry, Mississippi Department of Archives and History, Jackson, Miss.
[23] Barker, *With the Western Sharpshooters*, p. 11.
[24] OR series I, vol. 38, pt. 1, p. 126; also pt. 3, 415.
[25] Strayer and Baumgartner, *A Brief History of the 66th...*, pp. 112, 180.
[26] Barker, *With the Western Sharpshooters*, p. 39.

# BIBLIOGRAPHY

Barker, Lorenzo A., *With the Western Sharpshooters, Michigan Boys of Co. D, 66th Illinois*, reprint, Burlington, WV, 1994.

Baumann, Ken, *Arming the Suckers*, Dayton, Ohio, 1989.

Baumgartner, Richard A. and Larry M. Strayer, *A Brief History of the 66th Illinois Infantry, 1861–1865*, in a reprint of Lorenzo A. Barker, *With the Western Sharpshooters, Michigan Boys of Co. D, 66th Illinois*, Burlington, WV, 1994.
  *Echoes of Battle The Atlanta Campaign*, Huntington, WV, 1991.

Bowe, Prosper, 66th Illinois Inf., letter of July 28, 1864, Lonn Ashbrook collection, Bloomingdale, Mich., printed in Strayer and Baumgartner, *A Brief History of the 66th Illinois Infantry, 1861–1865*, in a reprint of Lorenzo A. Barker, *With the Western Sharpshooters, Michigan Boys of Co. D, 66th Illinois*, Burlington, WV, 1994.

Boyd, Capt. William S., document, 1864, in the possession of Dan Fagen, Florissant, MO.

Carlile, Richard, "The 1st District of Columbia Cavalry, Regimental History," *Miliary Imagaes Magazine*, Vol. VIII, No. 2, Sept./Oct. 1986, pp. 11–13, East Stroudsburg, PA.

Chambers, William P., 46th Mississippi Infantry, C.S.A., Journal Diary, 1864, Mississippi Department of Archives and History, Jackson, MS.

Dyer, Frederick H., *A Compendium of the War of the Rebellion*, New York, 1959, 4 vols.

Fagen, Dan P., "The Dimick Rifles of the 66th Illinos Infantry," *The Gun Report*, March 1996, Vol. 41, No. 10, p. 16 ff., Aledo, IL.
  "Swiss Chasseurs, Dimicks & the 66th Illinois Infantry," *North South Trader's Civil War*, Vol. 23, No. 2, March/April 1996, pp. 26 ff., Orange, VA.

Farries, Private William, 24th Wisconsin Inf., letter from Louisville, KY, Sept. 27, 1862, to "Dear Brother," author's collection.

Hamilton, John G., "The Dilemma of the Martially Marked Henry Rifle," *The Gun Report*, Vol. 32, No. 6, Nov. 1986, pp. 50–53, Aledo, IL.

*Henry's Patent Repeating Rifle Catalog*, New Haven Arms Co., New Haven, CT, 1865.

James D. Julia Auction Catalog, item no. 98 (Henry rifle no. 10294, 22" bbl.), April 26–28, 1999, Fairfield, ME.

*Louisville Journal* newspaper, June 26 to Dec. 31, 1862, Louisville, KY.

McAulay, John D., *Civil War Breech Loading Rifles*, Lincoln, RI, 1987.

Madaus, Howard M., "The First 1,550 Henry Rifles," unpublished paper, Cody, WY.

Marcot, Roy, *Spencer Repeating Firearms*, Irvine, CA, 1983.

Merrill, Samuel H., *The Campaigns of the 1st Maine and 1st D.C. Cavalry*, Portland, ME, 1866.

Mogelever, Jacob, *Death to Traitors, the Story of Gen. Lafayette C. Baker, Lincoln's Forgotten Secret Service Chief*, Doubleday Co., Garden City, NY, 1960.

New Haven Arms Company, Letter Book, May 1, 1857 to March 28, 1859, and October 8, 1862 to December 12, 1863, McCracken Research Library, Cody Firearms Museum, Buffalo Bill Historical Center, Cody, WY.

Parsons, John E., *The First Winchester, The Story of the 1866 Repeating Rifle*, New York, 1955, 1969, pp. 8–9.

Records of the Chief of Ordnance, Records Groups 74, 156, National Archives, Washington, D.C.

*The Scientific American*, new series, Vol. 8, Issue 10, March 7, 1863, pp. 150–151.

Thompson, C. Alexander, New Haven, CT, Jan. 27, 1863[4], letter to L. Dodge, author's collection.

U.S. House of Representatives, Document 89, 42nd Congress, 2nd Session, Serial No. 1511, Sales of Arms and Ordnance Stores, Washington, D.C.

*U.S. Martial Arms Collector and Springfield Research Service Newsletter*, No. 78, Oct. 1996, Springfield Research Service, Silver Spring, MD.

U.S. Ordnance Dept. Purchases, U.S. House Exec. Doc. 99, Serial 1338, Washington, D.C.

U.S. Senate Report No. 183, 42nd Congress, 2nd Session, Serial No. 1497, Sales of Arms by the Ordnance Department, Washington, D.C.

*War of the Rebellion: A Compilation of the Official Records of the Union and Confederate Armies*, 70 vols. in 128 parts, Governnment Printing Office, Washington, D.C., 1880–1901.

Williamson, Harold F., *Winchester, The Gun That Won the West*, New York, 1952.

Wilson, R.L., Winchester, *An American Legend*, New York, 1991.

[Winchester, Oliver F.] *Selection of an Arm for the Use of the Infantry and Cavalry of the United States Army, Being a Review of the Report of a Board of Officers Convened by the Secretary of War to Meet at Washington, March 10, 1866*, New Haven, Connecticut, 1868.

# INDEX

## A
A.C., 1st, 80, 81
Adams & Co., 28
Adams Express Co., 44
Albright, T.J. & Sons, 22
Allin, Lucian C., 51, 90
Ames, Lt. S., 83
Anderson, John, 82
Arcade Malleable Iron Co., 9, 25, 84
Arkansas, 1st, 78
Armstrong, Sgt. Gilbert, 83
Arny, Gov. W.F.M., 80
Atlanta Campaign, 25, 64–66
Augusta (Kentucky), 11

## B
Baker, C.M., 81
Baker, Col. Lafayette C., 28, 40, 43–46, 48, 51, 73
Baker, Maj. J.S., 31
Baker's Mounted Rangers, 45, 46
Ballard carbines, 47
Barker, Sgt. Len, 61, 62, 63, 66, 67, 80
Barnett, Paul, 83
Batchelder, John L., 82
Bates, R.H., 81
Beard, John R., 62
Beesley, Pvt. William P., 60, 67, 80
Benton, Maj. James G., 54, 93, 94
Bentonville, N.C. (Battle at), 40, 66, 70
Berdan, Col. Hiram, 57
Biggs, T.R., 79
Binger, F.W., 78
Bingham, Capt. H., 78
Birge's Western Sharpshooters, 17, 30, 57–70, 78, 80
*Black Hawk*, 27
Bliss, Col. Z.R., 83
Blunt, Maj. Gen. J.G., 80
Bodge, J.M., 67, 80
Bolton, D., 82
Boston (Massachusetts), 13, 22
Bowe, Pvt. Prosper O., 64
Bowen (Chicago retailer), 27
Bowen, E.R., 82
Bowin, Marvin D., 82
Boyd, Capt. William S., 63
Bragg, (Conf.) Gen. Braxton, 10, 11
Brass-framed rifles, 36, 66, 84, 85–88
Brown, John W., 9, 13, 13, 15–20, 22–25, 28, 44, 46, 73, 76, 78, 80, 85
Brown, Levi, 80
Brown, Lt. John, 28
Burkhardt, G., 81
Burton, Capt. J.H., 79
Buzard, M.G., 81

## C
Cairo (Illinois), 15
Cameron, Simon (Sec. of War), 41
Camp/Fort Davies, 58, 61
"Cat Gun," 39
Chapman, Charles G., 50
Chattanooga (Tennessee), 27
Cheney, Charles, 42
Child, Pratt & Fox, 12
Chipman, Col. N.P., 78
Clarksville (Tennessee), 11
Cleaveland, W.S., 24
Cleburne, Gen. Pat., 64
Cloudman, Maj. Joel W., 28, 30
Coffman, G.W., 83
Colt, Samuel, 15
Colt M1851 Navy Revolvers, 42
Colt's, 9, 15, 23, 42, 47, 54, 55, 84
Columbus (Ohio), 9, 12, 44, 46, 67, 76, 78
Conger, Lt. Col. E.J., 49
Conklin, J.H., 17
Connecticut, 1st, 79
Corinth (Mississippi), 58–61, 69, 70, 80, 90
Courtland, William, 11
Cowan, John, 80
Creasey, Lt. William J., 81
Crittenden & Tebbetts, 19
Cumberland, Army of, 25–26
Curtiss, Maj. D.S., 28

## D
D.C., 1st Cavalry, 28, 31, 40, 45–51, 54, 72, 73, 76, 79, 81, 83
Dahlgren, Capt. John A., 42
Dallas, Battle of, 64, 70
Damon, Milo N., 67, 79
Davies, Camp/Fort, 58, 61
Davies, John M., 32
DeLeuw, 82
Denee, Robert S., 78
DeRussy, Capt. G.D., 41

Dimick, Horace E. & Co., 22, 57, 58, 62
Dimick Rifles, 57, 59, 61–63, 66, 68, 69
Dodge, Brig. Gen. Grenville M., 62, 63
Dwight & Chapin, 47
Dyer, Maj. A.B., 51, 53, 54, 90, 92

# E

Ekstrand, John H., 27
Ellet, Brig. Gen. Alfred W., 28, 43
Ellithorpe, Lt. Col. A.C., 24
English & Atwater (England), 25

# F

Fahnestock, Capt. Allen L., 26, 79
Febiger, Maj. G.L., 79
Fox, John, 80
Freman, Lt. D.M., 81
French, Maj. Gen. Samuel G., 64
Fulton, G.W., 82

# G

Gardner, Dr. W.W., 15
Gleason, Thomas E., 67, 80
Government purchases, 28, 31, 43–49, 51, 55, 56, 73, 74

# H

Ham, Nicholas, 81
Hancock, Maj. Gen. Winfield S., 55, 83
Harris Light Artillery, 11
Hayden, Joel, 46
Henry, B. Tyler, 6, 8, 9, 31
Hoagland, Henry V., 80
Hodgson, Samuel, 36, 39
Holmes, A.C., 80, 83
Holway, Pvt. S.A., 81
Horton, Asahel, 67, 79
Hyman, Capt., 82

# I

Illinois  7th Infanry, 25, 63–65, 77, 78, 80–82
23rd Vol. Infantry, 28, 80
32nd Infanry, 30
36th Infantry, 27, 77, 81
51st Infantry, 27
64th Infantry, 25, 30, 81
66th Infantry, 25, 30, 35, 40, 57–70, 78–80
73rd Infantry, 27, 77, 81
86th Infantry, 26, 27, 79
96th Infantry, 27
97th Infantry, 80, 81
Indiana  65th Infantry, 27, 79
97th Infantry, 27
Iron-framed rifles, 24, 37, 42, 62, 69, 71, 84, 85–87

# J

Jackson, Sgt. W.J., 79
Juarez, Benito, 31

# K

Kautz, August V., 51
Kennesaw Mountain, 64
Kentucky, 1st C.S.A. Cavalry, 11
Kentucky, 12th U.S. Cavalry, 16, 17, 24, 25, 27, 62, 79, 82
Kentucky (State of), 16, 17, 27, 43
King, Nelson (frame loading patent), 31, 56
Kingsbury, Col. C.P., 41
Kittredge, B. & Co., 15, 16, 85
Kurtz, Col. John, 46

# L

Laidley Board, 53
Leader, J.M., 82
Lee, Col. William R., 46
Leets & Co., 51
Lightning Brigade, 24
Lincoln, Pres. Abraham, 43, 78
Litchfield, H.G., 32
Litherland, Pvt. David (Daniel?), 67, 80
Long, Col. Eli, 82
Louisville (Kentucky), 9–12, 16, 20, 22, 23, 62, 66, 78
Low, James & Co., 10
Lunt, W.F., 79

# M

Maine, 1st Cavalry, 28, 51
Marshall, J., 67, 78
Martin, E.B., 7
Martin, J.B., 80
Martin, Lt. Col. George A., 79

Massachusetts Arms Co., 22
Maxmilian, Emperor, 32
Mayfield (Kentucky), 13, 15
McAluster, Archd., 83
McClure, J.A., 81
McCook, Judge (Major) Dan, 10, 29, 79
McPherson, Maj. Gen. James B., 25
Meese, Frank W., 83
Meijo, Emperor, 83
Merrill, Capt. Edward J., 82
Merwin & Bray, 47, 73
Miller, Abraham, Jr., 83
Miller, J.C., 81
Mississippi, 46th Infantry, 64
Mississippi Marine Brigade, 28
Michigan, 66th (Co. D), 62
Mitchell, Lt. W., 42
Montross, C.A., 28
Moore, Amos F., 83
Moore, Sgt. E.A., 81
Morgan, John Hunt, 27, 29
Morris, A.W., 79
Morse, Henry A., 82
Mosby, John S., 28. 45, 46
Mount Zion Church (Missouri), 58, 70
Moyers, W.T., 78

# N

Naval Bureau of Ordnance, 42
Nelson, Pvt. Marcus S., 59, 60
New Haven Arms Company, 7–10, 15, 17, 22, 24, 42, 44, 45, 47, 52, 55, 84, 92

# O

Ohio, 66th Vol. Infantry, 23
   68th Infantry, 30
Orcutt, J.D., 81
Osborn, A.C., 79

# P

Padgett, David, 67, 79
Paducah (Kentucky), 15
Parish, C., 67, 79
Parkhurst, Henry, 82
Passavant, Walter, 82
Pauline, J.G.S./G.M.S., 80
Peace, Sgt. D.A., 64, 70
Pease, F.D., 82
Pennsylvania, 11th Cavalry, 30
Pennsylvania, 21st Infantry, 74
Pennsylvania, 98th Infantry, 74
Perryville, Battle of, 11
Petit, George W., 83
Pfeifer, T., 83
Philip Wilson & Co., 22
Phillips, John F., 80
*Pittsburgh* (U.S. gunboat), 27
Prentice, George D., 9–12, 16–18, 27, 44, 78
Prussia (Germany), 22
Putnam, Sgt. E.A., 83

# Q

Quinius, Louis, 67, 79

# R

Ramsay, Brig. Gen. G.D., 40, 44, 46, 48, 50, 51, 54, 92
Ramsey, James E., 79
Randall, Pvt. John, 64
Read, William & Sons, 22
Reed, Pvt. David, 81
Remington & Sons, 47
Rideout, Jacob, 83
Ripley, Brig. Gen. James M., 41–44, 46, 51
Rolf, A.J., 80
Rollins, W., 83
Root, E.K., 47
Rowland, Maj. Wm. S., 30

# S

Sanford, Lockwood, 81
Scabbards, Henry rifle, 34
Schick, Sgt. R.L., 83
Schuessler, J.W., 67, 78
*Scientific American, The*, 43
Semple, A.B. & Sons, 10, 12, 22, 23, 62
Serial numbers, 71, 75, 76, 85
Shaler, Brig. Gen. Alex., 82
Sharps rifles, 42, 55
Sharpshooters, 1st U.S., 30, 57
Sharpshooters (Birge's Western), 17, 30, 57–70, 78, 80
Sherman, Maj. Gen. Wm. T., 25, 66
Simison, S.A., 28
Sitting Bull, 82
Skelton, Wm. S., 78
Smith, J.A., 78
Smith, Maj. J.C., 83
Smith, James H., 80

Smith, Capt. John A., 83
Smith, (Conf.) Gen. Kirby, 10
Smith & Wesson, 8, 19
Smoot, Lt. Wm. S., 51, 90
Spangenberg, J., 82
Spencer carbine, 55, 91
Spencer rifle, 24, 27, 28, 41, 42, 54, 55, 65
St. Louis (Missouri), 12, 16, 22, 57–62, 66
Stanton, William C., 9, 10
Stanton, Edwin M., (Sec. of War), 43–45, 51, 78, 80
Stockton, Lt. Howard, 54, 55, 93, 94
Stockwell, L.C., 81
Stokes, John M. & Son, 10
Stony Creek (Virginia), 51
Stoughton, Edwin H., 45
Stout, H.C., 79
Stratton, Maj. Franklin A., 30
Swan, Maj. T.J., 79

# T

Tallman, L.P., 67, 80
Tennessee, Army of, 11, 25, 66
Tennessee, 5th Cavalry, 11, 78
Tennyson, John S., 27
Thomas, Maj. Gen. George H., 25
Thompson, Sgt. Albert C., 64
Thompson, S.H., 83
Thornton, Col. William A., 50
Titze, William C., 64, 70
Tyler, Enos W., 67, 79

# U

U.S. Vet. Vol. Infantry, 3rd, 41, 44–46, 55, 56, 74–76, 82, 83

# V

VanBrocklin, J.W., 67, 80
Venard, Stephen, 79
Vestal, S.W., 82
Vinson, Pvt. Hiram S., 67, 79
Volcanic Arms Co., 8

# W

Walch, John, 7, 9
Walch Revolvers, 9
Washington, D.C., 41, 43, 44, 46, 47, 50, 51, 54, 66
Washington, D.C. Arsenal, 44, 50, 51, 54
Washington Navy Yard, 42
Watson, J.K., 81
Watson, Peter H., 43, 46, 47, 81
Webster, Chas., 67, 80
Welles, Gideon, 42, 43
Werle, Pvt. Jacob, 74, 82
West Virginia, 10th Mounted Infantry, 30, 80, 83
14th Vol. Infantry, 30
Wheeler & Wilson Sewing Maching Co., 32
Wheeling (W. Virginia), 28
Wilder, John T., 24
William Read & Sons, 22
Williams, Judge R.K., 15
Wilson, Capt. James M., 16–18, 24, 27, 82
Wilson, Mathew, 23
Wilson, Philip & Co., 22
Wilson, Lt. William, 62
Winchester, Oliver, 7–13, 15–20, 22–25, 27, 28, 31, 32, 41–44, 46–48, 50, 51, 53–56, 71, 79, 83, 84
Winchester Repeating Arms Co., 6, 32
Woodward, Lt. Col. Thomas G., 11
Wright, S., 79

# Y

Yates, Richard, 58
Yerington, George W., 40, 79

# A Fine Selection of Civil War Titles

**CIVIL WAR SMALL ARMS OF THE U.S. NAVY AND MARINE CORPS**
by John D. McAulay. This book covers each of the war years in surprising detail, listing specific weapons used on specific ships and in specific engagements. The variety of weapons covered is amazing. Sailors and Marines fought in countless land battles, river actions and skirmishes where small arms played a crucial part in the action. 186 pp., 216 photos, 8.5"x11".
Hardcover • **$39.00 + p/h**

**BURNSIDE BREECH LOADING CARBINES**
by Edward A. Hull. This important volume presents a detailed analysis and a model-by-model study of this popular Civil War cavalry arm. This affordable book contains all kinds of information about Burnsides that is unavailable in any other source. Black & white photos, 95 pp., 7"x9.5". Hardcover • **$16.00 + p/h**

**SHARPSHOOTER: Hiram Berdan, His Famous Sharpshooters and their Sharps Rifles**
by Wiley Sword. The story of the inventor, his men and the innovative weapon they used. No collector of Civil War firearms will want to miss this book, which is written by one of the most famous and successful authors. 125 pp., 7"x9.5".
Softcover • **$18.00 + p/h**

**FIREPOWER FROM ABROAD**
**The Confederate Enfield and the LeMat Revolver**
The British-made Enfield rifle was so valued by the South during the Civil War that blockade runners cargoed shipments right up to the last days of the war. Another weapon with strong romantic and historical attachments to the South was the LeMat revolver. Chronicled on these pages is the story of just how these weapons reached the Confederate market — a story rich in international intrigue and details about the weapons themselves. As a bonus, information regarding a variety of other Confederate small arms is presented within the appendices. 120 pp., 7"x9.5".
Hardcover • **$23.00 + p/h**

**CIVIL WAR CARTRIDGE BOXES OF THE UNION INFANTRYMAN** by Paul Johnson
There were four patterns of infantry cartridge boxes used by Union forces during the Civil War. Quoting original Ordnance Department letters, the author describes the development and subsequent pattern changes to these cartridge boxes that were made throughout the rifle-musket percussion era. 175+ photos, 352 pp., 7"x10". Hardcover • **$45.00 + p/h**

**THE CONFEDERATE LeMAT REVOLVER**
This new book by Doug Adams describes LeMat's wartime adventures aboard blockade runners and alongside the famous leaders of the Confederacy, as well as exploring the unique revolvers that he manufactured for the "Southern Cause." Nearly 200 full-color photos! Over 70 B&W period photos, illustrations and patent drawings. 112 pages, 8.5"x11".
Deluxe Softcover • **$29.99 + p/h**

**AMERICAN SOCKET BAYONETS & SCABBARDS**
by Robert M. Reilly. The standard guide to socket bayonets, with hundreds of examples illustrated and explained. Covers from Colonial times through after the Civil War. This book is especially good at helping you to identify unknown examples. The illustrations show the markings and details that you need to look for. 208 pp., 9"x12". Hardcover • **$45.00 + p/h**

**CIVIL WAR PISTOLS OF THE UNION**
by John D. McAulay. Union handguns of the War between the States, including government procurement information, issue details and historical background. Also covered are Union pistols used by Confederates, with specific details. This is one of the hottest collecting fields today, and McAulay's guide is the standard reference. Black & white photos. 166 pp., 8.5"x11".
Softcover • **$24.00**   Hardcover • **$36.00 + p/h**

**CIVIL WAR ARMS MAKERS & THEIR CONTRACTS**
Huge reprint of the famous Holt-Owen Commission report on Ordnance and Ordnance Stores, 1862. Contains an unbelievable wealth of details about more than 100 manufacturers and suppliers of Union guns, pistols, swords, accoutrements and cannon. Over 600 pages! A maker's index helps you to find the information you need. 6"x9".   Hardcover • **$39.50 + p/h**

**CIVIL WAR ARMS PURCHASES AND DELIVERIES**
The single most quoted source of information about Civil War weapons. Many authors have called it the "bible" of Civil War arms research. Also known as "Executive Document #99," this massive publication is a comprehensive list of every weapons purchase made by the Union. Items covered include muskets, carbines, swords, pistols, ammunition, bayonets, cannon, accoutrements and more. Each entry lists the contractor's name, the size of the order, the date of the purchase, the price, the specifications of the weapon, the date of the original contract and the date of actual payment — all taken from the original government ledgers. 312 pps., 6"x9". Hardcover. **$39.50 + p/h**

**CARBINES OF THE U.S. CAVALRY, 1861–1905**
by John D. McAulay. This book covers the entire crucial period stretching from the beginning of the Civil War to the end of the cavalry carbine era in 1905. Includes the War between the States, the Indian Campaigns, Custer's Last Stand, the Rough Riders in Cuba, the Philippine Insurrection and the Boxer Rebellion — all illustrated with 127 photos. 144 pp., 8.5"x11".   Hardcover • **$35.00 + p/h**

**THE AMES SWORD CO., 1829–1935**
by John D. Hamilton. A comprehensive history of America's foremost sword manufacturer and arms supplier during the Civil War. In these pages, you will find the complete sweep of the company's production, from small arms to cannon and the finest swords ever produced. 255 pp., 8.5"x11". Color Section.
Hardcover • **$45.00 + p/h**

---

**SHIPPING: $4.50 PER BOOK, $1.50 EACH ADDITIONAL BOOK.**
Order from: Mowbray Publishing, 54 E. School St., Woonsocket, RI 02895
or Call toll-free at 1-800-999-4697 or (401) 597-5055 • Fax (401) 597-5056
**YOU CAN ORDER ONLINE: order@manatarmsbooks.com**